The Information Warfighter Exercise Wargame

Second Edition—Rulebook

Christopher Paul
Jonathan Welch
Ben Connable
Nate Rosenblatt
Jim McNeive

Prepared for the Marine Corps Information Operations Center
Approved for public release; distribution is unlimited.

For more information on this publication, visit **www.rand.org/t/TLA495-5**.

Published by the RAND Corporation, Santa Monica, Calif.

Cover: RAND photo by Stephen Webber.

Contents

Figures and Tables

Figures

Tables

The Information Warfare Exercise Wargame: Example of Play from a Player's Perspective

Twenty-two of us arrived at IWX last week and started receiving blocks of instruction on information capabilities and activities and on information as a Marine Corps warfighting function. Midweek, they told us that the whole second week would be a wargame and that we'd be developing a concept of information for our side in the game. They broke us into teams; I'm on the BLUE team, which means that I'm one of 11 marines playing as part of the information OPT for a Marine Expeditionary Unit (MEU). The other 11 participants in our IWX class are playing as the RED team, which means they're like an information OPT for the Centralian 17th Mechanized Infantry Battalion. The mission of the Combined Joint Task Force—of which the MEU is a part—is to land in Montanya (a U.S. partner nation recently invaded by Centralia); secure an airfield, a port, and the nearby city to protect the welfare of the Montanyan population; roll back Centralian forces; and enable follow-on operations.

We spent the last few days of the first week going over all the intelligence we had on Montanya and the Centralian forces, familiarizing ourselves with the MEU's planned concept of maneuver, and working on our concept of information. We finished up our plans on Friday, gave a confirmation brief, and started the wargame on Monday. I guess it is true what they say about no plan surviving contact with the enemy, because things have not been going our way. Right away, turn 1 included an inject that one of our forward companies had gotten pinned down at an intersection, lost several of their Light Armored Vehicles, and taken heavy casualties. That forced changes in the planned concept of maneuver—which EXCON took care of—but the inject also forced changes in how our concept of information related to the updated concept of maneuver. We scrambled and adjusted our planned actions to retarget locations where our marines actually were (rather than where they were supposed to be if they had advanced on schedule) and came up with a clever feint and supporting information efforts to relieve some of the pressure on the marines fighting at the intersection.

Now, it is Wednesday morning of the second week of IWX, and we've started turn 3. We just got the update brief from someone from EXCON (the player's guide I have calls the update briefing "Step 1"). As we feared, Regional Highway 2, the route toward the nearby city from the airfield we've secured, is congested by refugees. That's okay, I'm on it. We're now in Step 2, where we prepare actions to present, and it's my turn to present. This morning, I'm refining my briefing for our operations officer, the S3. The action I'll be briefing is a multi-capability effort designed to clear the route so the MEU can advance unimpeded while simultaneously directing local Montanyan civilians to a disaster relief site that the civil affairs folks are establishing on the other side of the airfield (near the civilian terminals and accessed by a different route than the one the MEU intends to use to advance).

Time is called to start Step 3, where we present actions for approval, and we all stand as the S3 role-player comes into the room. Since I'm briefing first, I start with a quick review of the situation and orient everyone to the BLUE team game board. I point out the airfield, where MEU elements are at or near the airfield, and the big blue arrow showing the intended line of advance. Then I point out the cluster of civilian icons along that route and note that this is the problem this action is intended to address. I also point out where the relevant information capabilities are: the icon for the civil affairs detachment and their relief center, the icons for the psychological operations (PSYOP) loudspeaker teams and the mobile broadcasting capability, and, back on the ship, the location of the production shop that will produce the leaflets I'm going to propose we drop.

With the stage set, I then describe how we're targeting two audiences: civilians who have evacuated and civilians who have remained at home along and adjacent to Regional Highway 2. Radio messages, loudspeaker broadcasts, handbills, and leaflets, all in the Montanyan language, will instruct civilians to shelter in place to avoid injury from fighting in the area, and will instruct those who cannot remain in place or those who need medical assistance to proceed to a displaced persons and aid site on the north side of the airport, with instructions regarding the roads to use to get there.

I finish by explaining that the desired end state is dramatically reduced traffic on Regional Highway 2, with a secondary effect of increased civilian awareness of and activity at the civil affairs relief site. I list our measures of performance (MOPs): number of leaflets dropped in the target zone, hours of radio broadcast, and hours of loudspeaker broadcast. I explain that our measures of effectiveness (MOEs) should be observable by reconnaissance assets or through overhead imagery, and also documented in reports from executing units: What do the loudspeaker detachments see happening when people hear their messages? What level of activity does the civil affairs detachment observe?

The S3 nods approvingly, and then asks if there is time to get all of these military information support operations (MISO) products approved.

It's a good question. I say that, yes sir, it is all good to go, as all the messaging is within the bounds of pre-approved MISO products and series. MISO always has a standard set of pre-approved series aimed at civilians: "stay" messages (shelter in place), "go" messages (evacuate, with instructions on where to evacuate), and messages about humanitarian relief and where to go to get it.

The S3 then asks me if I have my three reasons this action will succeed ready to go for the engagement step. I do have them ready, and I share them.

The S3 approves my action. After briefs from other team members, the S3 approves a second and third action as well, then departs. We have about 15 minutes left before we'll head to the engagement room for Step 4. We take turns trying to think of stuff the other team might say in their rebuttals, stuff about why our actions might not work, and things we might say in our counterarguments.

As we enter the engagement room, there is an excited hush. The engagement game board is a giant laminated plotter map that covers most of the floor. The judges sit behind a table on one side of the room, with the GREEN representative next to them on one side, and the timekeeper (who operates the big countdown clock) on the other. Opposite the judges sit the narration team and some other EXCON personnel. We, the BLUE team, sit behind a table along one of the remaining sides, with the RED team directly opposite us, completing a square.

The head judge calls engagement to order. The narrator gives a quick overview of the current location of both sides' forces and the current status of the operation. It is always interesting to see the hybrid view of both RED and BLUE situational awareness. I know that "what happens in the engagement room stays in the engagement room," but it is still interesting to see where the RED forces actually are and what their maneuver intent for the turn is.

The head judge calls the team leaders up to determine the order of actions. We'll present our actions in alternating sequence depending on who rolls highest. Our team lead rolls highest, so my action, action BLUE 1, is first.

I step "into the arena" and stand directly on the giant plotter map. The RED team presenter for their first action joins me and gives me a bit of a stare-down as we face off for the first round of presentations.

The timekeeper starts the clock, and it begins to count down from 5 minutes. As the clock ticks down, I lay out the details of the action pretty much the same as I did for the S3. I pace around the plotter map and use a pointer to show where the different capabilities will be and where the different drop zones and broadcast areas are. I finish well before my time expires with the three reasons this action will succeed. First, I note that the people are scared and confused and that these products offer trustworthy information and clear guidance and so should be effective. Second, I note that the two behaviors we're promoting (stay in place or head to a relief site) are probably things many members of the target audience were considering doing anyway, and our messaging just provides more information and instructions regarding that decision. Third, I note that the action will succeed because so many members of the target audience will get the message, sometimes repeatedly and through multiple media: We've got a bunch of different communication modes that we're employing.

My team lead flashes me a thumbs up. The timekeeper resets the clock to 4 minutes, and the head judge instructs the RED team to prepare their rebuttal. They circle up and begin to discuss furiously. We huddle up as well and discuss some of the possible lines of rebuttal the other team might use.

The timekeeper calls "time" and resets the clock to 2 minutes, beginning the rebuttal countdown. My RED counterpart in the arena offers three reasons they think the action will fail. First, she notes, the messages are all clearly from the U.S. force, and the United States is not a credible messenger in this area. Second, even if this action makes some people shelter in place and draws some away to the north side of the airfield, the general level of panic among the civilian population is still going to leave Highway 2 too congested to allow the MEU to advance. Third, she says, she and other teammates know that MISO messages take a long time to produce and approve; she asks how long we've been planning on this action, and if that includes enough time for product approval?

I stand to launch my counterargument, but before I start the timekeeper waves me back to the team and reminds me that we have 2 minutes to discuss our counterarguments before I make them. The timer starts ticking down from 2:00. It doesn't take us long. We discuss for less than a minute, and I then announce, "I'm ready." The timekeeper resets the clock to 1 minute, starts the timer, and says, "Proceed with counterarguments." I offer my counterarguments, noting that in fact our intelligence summary indicates that the United States is well liked and well respected in this part of Montanya, so messages from U.S. forces should be viewed as highly credible. And, even though this action is a relatively new development in our plans, based on observing displaced persons clogging roads, as I told the S3, MISO always has messages of this sort as part of their pre-approved materials.

The judges briefly confer, making notes on their scoresheets. The head judge asks the room, "Well, which is it? Is something like this part of a pre-approved series, or not?" Several of my teammates nod and say, "Absolutely," but it is firmly sealed when one of the other EXCON personnel, a MISO subject-matter expert, confirms that such products are typically pre-approved. One of the other judges asks about timelines for developing, printing, loading, and distributing leaflets. I don't know the answer, but one of my teammates does; I confer with him for a moment, then share his observations with the room, and his timeline is confirmed by that same MISO expert. The head judge turns to the GREEN representative on EXCON. The GREEN rep says he doesn't have any questions but that he wants to confirm that the United States is well regarded in this region of Montanya, so messages from the USMC should not be viewed as inherently lacking credibility.

The judges confer and work on their scoresheets, and we talk quietly to each other. The debate seems like it went our way, and so did the judges' questions and observations. Maybe we'll get a nice, low target number.

The judges complete their scoresheet, and the head judge signals for attention. The head judge announces that this action is fairly straightforward to execute, so the target number for execution is 6. The whole team is pleased. Six is the lowest target number we've had for any action so far, so we have a really good chance that the action is executed successfully.

I step forward toward the dice. My action, my chance to roll. I roll all three dice. They come up 4, 4, 1—a total of 9. Nine is not that great a roll, but it is good enough here, as it beats the target number by 3. The head judge announces that the action has been successfully executed, which means that I get to roll for its effectiveness.

The head judge announces that the target number for effectiveness will be 9. Not as easy as the 6, but still a pretty decent shot. I pick up the dice and roll again. 4, 4, 4—a total of 12 this time! Again the roll beats the target number by 3, and the head judge announces, "Full success." The narrator begins to describe the outcome of the action: All the different messaging products go out without a hitch, the messages are well received, and the target audiences begin to behave as we hoped. We won't know the full story of what happens until Step 5, results and reset, as maybe one of RED's actions affects one of those audiences or affects something else in the battlespace that prevents us from fully accomplishing our objectives—but the action, my action, was a success!

Step 4 proceeds with a RED presentation, which I have to rebut. After that, my time in the arena is done, so I get back behind our team's table to help support our other presenters. Our other actions and one of RED's remaining actions fail, but RED gets a partial success with a fake atrocity video, and the narrator says that this leads to growing indignation with the U.S. effort both back in Centralia and in the international community. I'm already starting to think of things we might do to help head that off.

In Step 5, the narrator summarizes all of the consequences of the actions and wraps up the maneuver progress for the turn. Because of our successful action, the route of advance is clear, and the MEU is able to advance into the outskirts of the city, meeting the commander's objective for the turn, but it is clear that the atrocity video is going to demand our attention next turn.

1. Introduction to the Information Warfighter Exercise Wargame

Welcome to the Information Warfighter Exercise (IWX) Wargame rules. To get a quick start on understanding the rules and the order of play, proceed directly to Section 1.4.

The Marine Corps Information Operations Center (MCIOC) conducts an IWX two to three times per year. Prior to 2020, many IWXs (some under the prior exercise nomenclature Combined Unit Exercise, or CUX) included a wargame-like "opposed free play" element in which teams of participants prepared plans for information activities and then revised them based on friction encountered within a notional scenario and perhaps as opposed by other teams. In these prior iterations, the wargame-like exercise elements lacked a structured adjudication mechanism, which both trainers and participants found unsatisfying. For the 2020 IWX cycle, MCIOC asked RAND to help in developing a more structured wargame with a formal adjudication process. The ruleset that was developed, playtested, and implemented through that effort was published in 2021 as *The Information Warfighter Exercise Wargame Rulebook*. The original rules provided the foundation for several years of IWX Wargames, which provided feedback on how the game rules could be changed, expanded on, and refined. RAND published a supplement to the rules that added guidance and rules for conducting the IWX Wargame at the operational, rather than the tactical, level of war and for running IWX Wargame scenarios that take place during competition rather than conflict. This document, *The Information Warfighter Exercise Wargame: Second Edition—Rulebook*, integrates the supplement rules into the core rules for the game, updates the ruleset based on experiences and experimentation with the wargame over three full years of IWX cycles, expands the scope of elements considered as part of the wargame, and updates the rules to be consistent with the language of current Marine Corps doctrine for information.

The IWX Wargame involves two teams of players competing against each other in and through the information environment to better support their respective sides in a notional scenario. The rules for this wargame are scenario-agnostic and can be used with any appropriate scenario at any level of classification. See Section 3.4 for scenario requirements.

Teams represent an information working group (IWG) or information operational planning team (OPT), or its adversary force equivalent, as dictated by the scenario. Each team is responsible for generating a plan for information activities as part of the information warfighting function's contribution to a larger plan to meet the objectives of the force that the team is part of. The larger plan for the operation, including the overall commander's intent and concept of maneuver, is largely fixed as part of the scenario and is part of the planning context.

Teams of players should bring a well-developed concept of information to the wargame from the prior week of IWX training (the wargame is intended to take place during the second week of a two-week IWX). Playtesting and experience have revealed that a planning period prior to the start of the wargame makes the game much more successful. As part of the wargame, players will be called on to add details to their plan, amend their plan dynamically in response to in-game events, prepare discrete game actions as part of plan execution, and make cogent arguments in favor of their team's actions and against the actions of the opposing team.

OPTIONAL: Instead of taking place during a two-week IWX, the wargame might take place during a single-week IWX, with the planning period considerably compressed or executed partially before participants arrive at IWX (including read-ahead and perhaps remote presentations or discussion) and the number of turns of the wargame reduced. See Section 4.2.

1.1. Using This Rulebook and Associated Materials

Sections in these rules are numbered for easy reference and cross-reference. The table of contents is a useful reference for finding rules regarding specific topics.

- Section 1 is the introduction and discusses the goals of the game, the terms used throughout the ruleset, and the overall flow of the game from turn to turn.
- Section 2 details each of the five steps of a turn within the wargame and is the most extensive section within the ruleset.
- Section 3 provides guidance for preparing to conduct the wargame—including staffing and roles required to conduct the game, details about scenario preparation (the wargame rules are scenario-agnostic, but a scenario is required), preparing the physical spaces, game boards or maps, and representations or tokens for the game—and estimates of the time required for each step within each turn.
  - Section 3 concludes with considerations related to whether the game is being run by an expeditionary cadre rather than at the game organizers' home station.
- Section 4 provides optional rules related to increasing or decreasing the scale of the game, either by adding additional participants or by reducing the time allotted to the wargame.
- Section 5 provides optional rules for conducting the wargame at the operational (rather than tactical) level, or for a competition rather than a conflict scenario.

1.1.1. Optional Rules

Throughout this document, some rules—such as the one on the previous page, about a one-week IWX—are flagged as OPTIONAL and set in blue boxes. Any of these rules may be included in any IWX Wargame at the option of the trainers running the game. As a practical matter, the lead of the group of exercise controllers will have the final say regarding which optional rules are in play and which are not.

1.1.2. Downloadable Player's Guide

This rulebook is intended for those running the IWX Wargame. A brief *Player's Guide* is available at www.rand.org/t/TLA495-5, and players should be given that document rather than this full rulebook (lest better understanding of the nuts and bolts of the game lead them to "game the game"). Players can be sufficiently prepared to play through verbal instruction and the *Player's Guide.*

1.1.3. Downloadable Game Aids

Various scoresheets, checklists, and other printable game aids (GAs) are available at www.rand.org/t/TLA495-5. In this rulebook, we refer to these by number and name—for example, *GA1: Turn Structure Overview* and *GA3: Judge's Responsibilities.* Table 1 lists the name and number of each game aid.

TABLE 1
Downloadable Game Aids

#	Name
GA1	Turn Structure Overview
GA2	EXCON Lead's "Run of Show" Summary
GA3	Judge's Responsibilities
GA4	Step 4 Sequence of Play
GA5	S3 Role-Player's Responsibilities
GA6	S3 Role-Player's Action Approval Checklist
GA7	Narrator's Preparation Worksheet
GA8	Engagement Scoresheet
GA8a	Single-Roll Engagement Scoresheet
GA9	Guidance for Pre-Approval Panelists
GA10	Making an Outcome Roll

1.2. IWX Wargame Learning Objectives

1.2.1. IWX Overall Goal

The overall goal of IWX is to train and educate participants on **how to affect the decisions and behaviors** of relevant actors. This goal is primarily met through training and practice related to planning to influence target audiences at the tactical level.

The information warfighting function encompasses a range of capabilities, activities, and forces. IWX and the IWX Wargame places primary emphasis on the "influence capability and activity" category, with secondary emphasis on the "deception" and "inform" categories. Sometimes, depending on the mix of participants, other capability and activity categories might be included, such as "cyberspace" or "electromagnetic spectrum." It might be possible to include other Marine Corps information capabilities or activities in a given instance of the game, but doing so might require some improvisation or adjustments to the rules. **IWX (and this wargame) is intended to focus on efforts to influence, inform, or deceive in support of operations at the tactical level.**[1]

1.2.2. IWX Secondary Goals

Secondary goals of IWX include participants continuing to develop information planning skills that will aid them when they return to their home units and when they participate in real OPTs or IWGs in the future. Specifically, a goal of IWX is that participants will improve their presentation and briefing skills and their ability to articulate and defend their concept of information and the associated information activities.[2]

Other secondary goals relate to the overall atmosphere developed in IWX and the wargame. The atmosphere of the wargame is intended to foster the exchange of ideas and the sharing of tactics, techniques, and procedures. The exercise and wargame are intended to stress professional knowledge and promote further professional development. IWX is intended to promote an environment of interoperability across and between services, relevant interagency elements, and international partners. The IWX Wargame intends to do all these things while remaining a stimulating, exciting, and fun challenge for participants.

1.2.3. Objectives of the IWX Wargame

The IWX Wargame supports the goals described in Sections 1.2.1 and 1.2.2 by allowing play in an environment in which IWX participants execute their concept of information within the confines of a realistic scenario and against an opposed team that is seeking to execute its own concept of information. Both sides will be called on to adjust their plan to conform to realistic obstacles, to respond to the actions of their counterparts, or to otherwise react to situations that arise within the game and scenario that may be outside of the scope of their original plan.

During the game, success, failure, and mixed results will all be possible. The effectiveness of player planning and actions will be determined by the quality of the plans prepared by each side; their respective ability to present actions based on those plans; friction between opposing plans; friction between actions and the scenario environment; expert adjudication; and chance. In order to succeed, and as part of the game, teams of players must

1. Clearly describe their target, target audience (TA), or audience and the desired effect on that audience.[3]
2. Continually reassess the information environment and contend with unavoidable ambiguity in the information environment.
3. Articulate how selected information capabilities are to be deployed and offer reasons for the effectiveness of their employment.

[1] The game can also be played to emphasize planning for information at the operational level, but doing so requires some modification. See Section 5.

[2] The Concept of Information construct was introduced in 2024 in Marine Corps Warfighting Publication (MCWP) 8-10, *Information in Marine Corps Operations*. The latest version (2020) of MCWP 5-10, *Marine Corps Planning Process*, provides a template for Annex I, "Information," which will likely be replaced by the Concept of Information at the next update of that doctrine.

[3] Sometimes the subject of an action should not be referred to as a *target audience*, such as when seeking to protect the friendly force or the American public from disinformation or manipulation.

4. Explain how proposed actions affect the target, TA, or audience and how that effect supports their side's concept of maneuver and overall mission success.
5. Verbally present their game actions in a clear, concise, and well-thought-out statement with appropriate supporting justifications.
6. Defend the argument for the success of their actions against the counterarguments of their opponent, if required.
7. Present verbal counterarguments to their opponents' presented actions.

1.3. Terms

This section presents and defines the terms used throughout the ruleset. Terms are listed here and used consistently throughout to avoid confusion. Terms are not presented alphabetically but rather are aggregated by topic area (people in the game, structure and sequence, presentation and adjudication) and sorted within topic area based on primacy.

1.3.1. People in the Game

Participant: Someone attending IWX as part of the training audience.

Player: A participant at IWX who is playing the wargame. Players and participants should entirely overlap unless a participant is excluded from the wargame for some reason (such as only being present for a portion of IWX).

Side: The side in the conflict, in the wargame scenario, for which the teams play.

Team: The group of players representing an information OPT, IWG, or equivalent for one of the two sides in the wargame scenario. There are two sides and therefore two teams. Ideally, each team should have no more than 12 players and no fewer than 6. Each team represents an information OPT or IWG and only influences the plans and actions of their side in the conflict so far as they realistically would be able to as that body.

BLUE: The side that is playing the U.S. Joint Force, Marine Corps force, or allied or partner-nation force in the scenario.

RED: The side that is opposing BLUE objectives and pursuing its own objectives. For this wargame, players for the RED side will still use U.S. Marine Corps and/or joint planning doctrine and templates to produce their plans and actions.

Exercise Control (EXCON): The adjudicators, judges, and scenario managers responsible for maintaining all elements of the game environment that are not directly under the control of the players.

FIGURE 1
People in the Game

Photo credit: Jonathan Welch, RAND.

EXCON will be drawn from MCIOC personnel and other subject-matter experts (SMEs) as available and required. EXCON controls the maneuver elements of both BLUE and RED and all HQ elements (at all echelons) other than the OPTs represented by the player teams. EXCON also controls any other groups or forces in the scenario context that might otherwise be a separate exercise cell (GREEN, for example).

WHITE: In wargaming circles, EXCON is sometimes referred to as *WHITE*.

GREEN: Relevant noncombatant individuals and other groups and actors within the scenario. GREEN is not a separate team and is controlled by EXCON. EXCON may include an individual specifically responsible for planning and tracking GREEN actions or response, or this may be a general EXCON task.

Role-player: A member of EXCON who portrays a member of the staff of the RED or BLUE side outside of the teams. The wargame requires, for example, a role-player to play the operations officer (S3), to whom the teams brief their intended actions during Step 3 of each turn. Other role-players could be included depending on the scenario.

Narrator: A member of EXCON who describes (narrates) both (1) the performance and effectiveness of actions based on adjudicated outcome rolls and (2) the overall progress of the operation at the end of each turn. The narrator tells the story of the game as it unfolds from turn to turn. Some of that story is pre-scripted storyline within the scenario and based on maneuver plans, and some of that story is dynamically determined by the actions taken by the teams and the success or failure of those actions.

Target audience (TA): Throughout this ruleset, the subject of an action or proposed action is referred to as the *target audience*. However, there are certain circumstances where it would be inappropriate to refer to the subject of an action in that way—for example, when considering actions to defend the friendly force or the American public from adversary disinformation or communications disruption. In such situations, players are urged to use appropriate nomenclature (*audience*, *public*, etc.), and EXCON members are encouraged to mentally make appropriate substitutions within the rules (considering an audience or protected formation where a scoresheet calls for a "TA," for example).

1.3.2. Structure and Sequence

Turn: A cycle through the sequence of steps in the game. A turn includes receipt of an update about the scenario environment and progress of the operation; preparation and approval of multiple actions; engagement with the other team and discussion of all actions; and adjudication of the actions. The IWX Wargame typically consists of five turns, and each turn takes about 8 hours to complete (turns can be split across two half-days, so an afternoon and then the following morning, rather than being confined to a single day). Turns represent periods of time within the game scenario, but the amount of time represented by a turn may vary throughout the game. For example, turns when the tempo of maneuver forces is relatively low might represent longer periods of time.

Step: Each turn consists of five steps: (1) receive scenario and situation update, (2) prepare to present, (3) present actions for approval, (4) engagement and matrix debate, and (5) results and reset. Steps 1–3 are performed in spaces specific to each team, whereas Steps 4–5 take place in a shared engagement space.

Action: One of the building blocks of a turn. Within each turn, each team will prepare and present a number of actions (typically three). Each action must have a target, TA, or audience and a desired effect. An action can include multiple information capabilities in combination. Actions are argued for and against by the two teams during the engagement step. Actions are named and referred to by EXCON based on the team that presents them and the order in which they are presented, so action RED 2 would be the second action presented by RED in a given turn.

Continuing action/recurring action: An action that is begun in one turn and then continues on into later turns. Continuing actions are recognized as such by EXCON and may be subject to special adjudication. Continuing actions still need to be presented for (continuing) approval and presented and argued for and against in the engagement step.

Scenario: The notional but realistic context and operations that have brought the two sides into competition or conflict and provide the stage and field of operations for the wargame.

Storyline: How the overall scenario is expected to unfold except for the information parts. Essentially, the scenario will follow a preset script unless the teams' actions have a significant impact on progress toward one or more of the operational objectives. Even with the impact from information activities, the storyline is expected to remain within certain limits.

Phase: Phases within a given game are defined by the scenario and the overall concept of maneuver for each side that each team represents. Phases might include, for example, shaping, seizing the initiative, and decisive action. Individual phases in the operation depicted in the scenario will be pursued over one or more turns. Figure 2 shows a depiction of phases and turns in a notional game.

Inject: A change in the scenario introduced by EXCON during the game. Injects may be used to surprise or challenge the players, or to restore competitive balance if one team is doing exceedingly well or exceedingly poorly early on in the game. In some games, injects might not be used at all.

1.3.3. Materials and Setup

Game board or map: A display that represents the operational environment. Depending on the details of the execution of the game, the game board may be an actual game board, a large plotter-printed map, or an electronic map. Whatever its format, the game board will include representations of the line of advance or progress of the maneuver aspects of the scenario operation, and it will provide opportunities for the two teams to geolocate their information capabilities and their targets, TAs, and audiences. During Steps 1–3, each competing team will work from a separate board in the team's own planning space. EXCON will maintain a board that contains ground truth for the position and disposition of all forces. EXCON will also prepare an engagement board to be present in the engagement room during Steps 4 and 5 that will depict some hybrid of RED and BLUE situational awareness and that will be used for matrix debate and adjudication.

Game aid (GA): Various scoresheets, checklists, and other printable game materials are available for

FIGURE 2
Structure and Sequence of the IWX Wargame with Notional Phasing

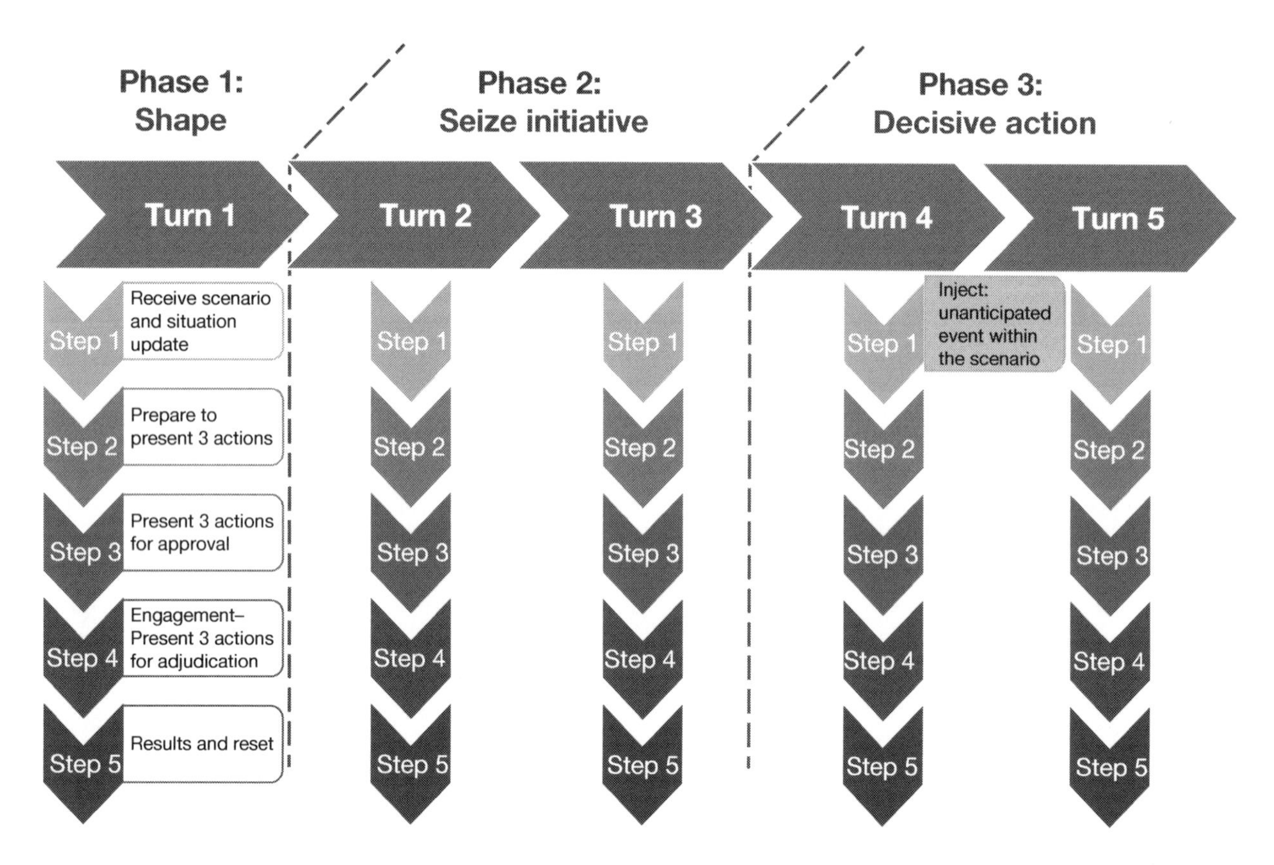

download at www.rand.org/t/TLA495-5, and in this rulebook we refer to them by number and name—for example, *GA1: Turn Structure Overview* and *GA3: Judge's Responsibilities*.

Planning space: A room or other designated space in which team members can plan without the other team overhearing. RED and BLUE will each have their own planning space.

Engagement space: A room or other designated space in which the RED and BLUE teams meet for engagement and matrix debate (Step 4).

Control space/control room: A room or other designated space for EXCON to operate in.

1.3.4. Presentation and Adjudication

Matrix system/matrix adjudication: A gaming system used to determine the success or failure of a given action. In a matrix wargame, adjudication centers on presentation and two-sided debate rather than on dice rolling. Dice rolls are used to help determine outcomes, but the effectiveness of each team's planning, presentation, and debate with the opposing team are the focus of the wargame and are important determinants of outcomes.

Presentation: Description of an action to begin adjudication, whereby the presenting player shares task, purpose, method, end state, and three reasons the action will succeed.

Presenting player: The player who is presenting an action. Players presenting for a team must be rotated, and the presenting player is the only one allowed to present arguments related to the action being discussed.

Rebuttal/rebut: Following the presentation of an action and a short discussion period, a single player from the non-presenting team offers up to three reasons that the action will not succeed. This is a presentation of reasons, *not* a counteraction. Rebuttal reasons may not include things that the non-presenting team would like to do to counter the action *unless* their approved actions or plans include a battle drill or standing operating procedure that would be triggered by the action.

Counterarguments: After the rebuttal, the presenting player may offer counterarguments. Counterarguments are limited in scope to being responses to the rebuttals from the non-presenting team. New points or issues may not be raised; only disagreements with or refutations of the rebuttal arguments are permitted.

Dice: Cubes with numbers from 1 to 6 represented on each face. In this wargame, three dice are rolled and added together. This will produce a result between 3 and 18 on each roll of the dice.

Outcome roll: For any adjudication event that has any element of chance, dice will help determine the outcome. The presenting player will roll three dice, add the results (for a total from 3 to 18), and compare the total with the target number set to accomplish the task effectively. This determines the outcome of the action or event. See Figure 3.

Target number: The number that a team is trying to reach or beat with an outcome roll. Target numbers are determined by EXCON and are a product of some of the events in the engagement step. If the total on the three dice is equal to or greater than the target number, then the action succeeds. If the total on the dice is less than the target number, the action fails. The distance between the target number and the actual roll determines the degree of success or failure.

Degree of success/failure: Some actions or events might have binary results: outright success or outright failure. Other actions—to be determined by EXCON—will result in varying degrees of success or failure. For these nonbinary events, the distance between the outcome roll and the target number indicates the degree of success or failure. For example, if the outcome roll is 14 and the target number is 11, then the degree of success is +3 (14 – 11 = 3), or "success by 3." These numerical values for degree of success will have corresponding effects in the scenario environment and the overall adjudication and storyline produced by EXCON.

Reroll: Some optional rules allow for failed outcome rolls to be rerolled under certain circumstances. This is what it sounds like: a mulligan, a do-over. Some caveats apply: (1) When a team chooses to use a reroll, the new result is final; if it is worse than the initial roll, too bad. (2) When using a reroll, all three dice are rerolled (not just one or a subset). (3) Rerolls can only be used on rolls made by the team using a reroll—that is, you cannot use your reroll to force the other team to reroll their outcome roll.

NOTE: The term *reroll* is not meant to apply to dice that fall strangely, such as when one falls on the floor or one leans against another die or other object and does not lay flat when the dice stop moving. When dice are thrown improperly (landing out of bounds or unsettled), they should not be read and should instead be thrown again. This does not count as a reroll, as the first roll was never properly completed.

FIGURE 3

Making an Outcome Roll

We roll three six-sided dice in this game to determine the success of an action. The dice are meant to simulate chance and factors beyond the control of the players and their role-played command staff.

Outcome Roll Procedure:

1. EXCON announces a *target number*—the number that you must equal or exceed on the outcome roll in order to succeed. Target numbers will vary based on the difficulty of the action, the quality of planning, and the Step 4 discussion.
2. Roll the dice!
3. Add up the three dice to get your outcome roll total.
4. If your total roll meets or exceeds the target number, you have succeeded! If it is less than the target number, your action has failed. (Depending on the specific game rules laid out by EXCON, you might have the opportunity to reroll.)
5. EXCON determines the degree of your success or failure based on the difference between the target number and your outcome roll total.

Example:

During engagement (Step 4 in the wargame) a player from BLUE presents an action. After hearing the presentation and matrix discussion, the EXCON judges determine that the target number for the outcome roll is 12. The presenting player rolls the three dice and gets:

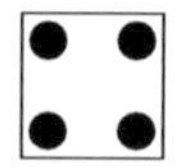

The dice add up to 15. This is greater than the target number (12), so the action succeeds!

The margin of success for this outcome roll is +3 because the roll total exceeded the target number by 3. EXCON will determine what this margin of success means in terms of the game narrative, depending on the action type and circumstances.

Tip: You have about a 91% chance of getting 7 or higher, about a 50% chance of getting 11 or higher, and about a 16% chance of getting 14 or higher.

1.4. Overview of Play

This section describes the sequence of play and provides a general outline of the conduct of the various steps of a game turn. These steps are also summarized in Figure 4 and in *GA1: Turn Structure Overview*. Section 2 provides much greater detail on each step.

The standard IWX Wargame consists of five turns (other instances may involve a different number of turns, but the steps composing each turn would remain unchanged regardless of the total number of turns). Each turn typically takes a full day to complete, but may be split over two half-days, beginning in an afternoon and completing the next morning. See Section 3.6 for a discussion of estimates of time required for each step within a turn. During each turn, both sides execute all steps. Turns are simultaneous. Teams alternate in presenting actions during Step 4, engagement.

Each turn consists of five steps:

1. Receive scenario and situation update
2. Prepare to present
3. Present actions for approval
4. Engagement and matrix debate
5. Results and reset.

Each step is briefly described in the following sections and described in much greater detail later in this ruleset. *GA2: EXCON Lead's "Run of Show" Summary* provides a high-level summary of all EXCON and player activities across all five steps of a turn.

FIGURE 4
Turn Structure Overview

	Player Activities	The IWX Wargame: Turn Structure Overview	EXCON Activities
		Step 1: Receive scenario and situation update	
Planning Spaces	Listen and take notes	Teams are updated on: » Intended progress and rough timeline » Maneuver plans for the turn » Positions of available capabilities » Note capabilities exhausted or attrited » Map changes » Status of GREEN » Injects	Set game boards Give situation update brief to teams
		Step 2: Prepare to present	
Planning Spaces	Plan/choose actions Identify order of presentation Pick presenters Prepare presentations Develop battle drills	Each presentation should include: » TA/audience » Desired effect » Purpose/connection to commander's intent » Conducting capabilities » Desired end state » Assessment plan » 3 reasons the action will succeed	Observe team preparations S3s prepare to rule on Step 3 proposals Complete RFI matrix as needed
		Step 3: Present actions for approval	
Engagement Space	Present actions to S3s for approval Finalize presentations of approved actions Prepare for rebuttal	» S3 approves or disapproves proposed actions » Roll for higher-level approval (if needed) » EXCON preps for Step 4	S3s review and approve/disapprove actions in consultation with other EXCON as appropriate S3s and observers brief judges Judges prepare draft scorecards
		Step 4: Engagement and matrix debate	
Engagement Space	Present actions Rebut other team's actions with 3 reasons why they will fail Counterargue rebuttals Make outcome rolls	For each action: » Presenting team presents *(5 mins)* » Other team prepares, rebuts *(4+2 mins)* » Presenter counterargues *(2+1 mins)* » EXCON questions, scores » Outcome roll and result	Set game board Reorient situation Set order of presentation Judges finalize action scores Judges meet to set target numbers Narrate execution and preliminary effectiveness
		Step 5: Results and reset	
Engagement Space	Take notes Plot to sustain success or seek revenge	» Summary of all results » Share operation storyline to date » In-stride AAR	Narrate full effects of each action and overall turn results Update storyline Prepare for the next turn Conduct in-stride AAR

GA1

1.4.1. Step 1: Receive Scenario and Situation Update (Overview)

Step 1 takes place in each team's planning space. During this step, teams receive an update on the current state of the operation; how (if at all) it is deviating from the planned concept of maneuver for their side; an estimate of the time in the scenario that will be covered by the turn; the current locations of available information capabilities (and notice of any capabilities that have been exhausted or attrited); the status of GREEN; and any updates on injects or changing scenario conditions. The Step 1 briefing summarizes the current state and reveals each commander's end state for the next phase of the operation (covering the next turn in the game).

1.4.2. Step 2: Prepare to Present (Overview)

Still in their planning spaces, teams revisit their plans as needed and prepare their actions for the turn. For each action a team wants to take, it prepares a presentation that will be given during Step 3 (present actions for approval) to a member of EXCON who will role-play the team's operations officer (S3). This presentation should include arguments in support of the action, which the team will also use during Step 4 (engagement). The presenting player for each action rehearses their presentation and makes sure they are ready to offer compelling explanations and arguments, both to the S3 during Step 3 and to EXCON in general during Step 4. This is also the step in which a team should identify non-organic assets, or permissions for special capabilities that their actions might require and will require additional approval in Step 3 from their S3 role-player. Any questions about available capabilities or permissions should be addressed to EXCON.

1.4.3. Step 3: Present Actions for Approval (Overview)

In their planning spaces, for each proposed action, the presenting player (and only the presenting player) briefs the S3 role-player on the audience and intended effects, capabilities involved, and details of (planned) execution. A different presenting player presents each proposed action. The S3 role-player approves or disapproves each action. (If any of the approved actions require capabilities or permissions not organic to the team's side's forces, EXCON will immediately adjudicate whether the action can proceed using a special outcome roll.)

At the end of Step 3, the S3 role-player will leave the team's planning space to back-brief EXCON judges on the approved actions, so that the judges can review them and begin preparing action scoresheets for use in Step 4.

1.4.4. Step 4: Engagement and Matrix Debate (Overview)

For this step, all players from both teams come together in the engagement space. Teams alternate presenting actions approved by their S3 role-player in Step 3. The same presenting player who presented actions in Step 3 presents them in Step 4. During engagement, each presenting player also presents three reasons why they believe their action will be successful. The opposing team then has a few minutes to huddle and identify three reasons why they believe the action will be unsuccessful or less effective than the presenting player has indicated, which they present publicly to the entire engagement room. Next, the presenting team has two minutes to prepare a response to their adversary's critique, which should not last more than one minute, and can include up to three counterarguments to the rebuttal. (This is a form of what is called *matrix adjudication* in the wargaming community.)

EXCON then has an opportunity to ask concise clarifying questions (but should avoid lecturing briefers unless the EXCON lead deems it necessary to take a training timeout). EXCON judges then quickly confer and finalize scoresheets, giving the presenting player a target number for the execution of the action, which reflects the judges' collective expert assessment of how difficult it will be to *execute* the proposed action. (This is distinct from its likelihood of achieving its intended effects, which is addressed in the second roll—think measure of performance [MOP]).

The presenting player then rolls three dice for an outcome roll, which produces a degree of success or failure for the execution of the action. If the action is executed at least partially successfully, EXCON gives the presenting player a target number, again assessed by the judges using scoresheets and a brief discussion to quickly come to consensus. The second target number reflects the likelihood of the action achieving its desired effects on its intended target audience (actions which are not at least partially successful in their execution do *not* move on to a second roll). The presenting player makes a second outcome roll, this time to determine the impact and effectiveness of the action. The narrator gives a preliminary description of how the action unfolds and its initial impact; the overall outcome of the action will not be narrated until Step 5, after all actions have been completed and adjudicated.

Teams alternate presenting actions, engaging in matrix discussion, receiving EXCON input and target numbers, and completing outcome rolls until all actions have been adjudicated.

1.4.5. Step 5: Results and Reset (Overview)

EXCON completes Step 5, recording all outcomes; narrating the results, consequences, and effects from all actions; and describing their impacts on the progress of the overall operation. EXCON makes any adjustments necessary to the overall storyline and to materials that will be presented as part of the update in Step 1 of the next turn. Step 5 concludes with an in-stride after-action report, during which EXCON staff may offer guidance and advice, or exercise personnel can take advantage of any teachable moments that emerged during the course of the turn.

Steps 1–5 are repeated for all six game turns. After the final step of the final turn, EXCON announces the final outcome of the scenario operation, including which team "won" (and why); presents any awards; and leads a more extensive after-action review (AAR) to cement lessons learned and to seek input to improve the wargame (and IWX) in the next iteration.

1.5. Inputs and Activities Required Prior to the First Turn

Details on preparing this wargame, including scenario preparation and the selection of EXCON personnel to run the game, are provided in Section 3 of this rulebook. We present the rules of the game first, in Section 2, and in that section we assume that the following things have happened as part of IWX prior to the wargame taking place:

- Participants have been divided into two teams of 6–12 players, one to represent the information OPT/IWG for BLUE and the other to represent the equivalent for RED.
- EXCON members have been chosen, mastered the rules of the game, and either prepared a scenario or sufficiently familiarized themselves with an existing scenario. There is some flexibility in how many EXCON members there need to be, depending on factors such as the scenario and whether any EXCON members are serving multiple roles. Typically, EXCON will consist of the following:
  - three judges, one of whom should be "EXCON lead"
  - an S3 role-player for each team
  - the narrator and scenario team, which should consist of
    - the narrator (who will describe the outcome of adjudicated actions and build results into the scenario storyline)
    - a notetaker assigned to each team
    - someone to manage the game board (the "map master")
    - someone with responsibility for keeping track of the facts of the scenario (the "reality master")
  - an action officer who wrangles participants, determines the length of breaks, etc.
  - individuals with responsibility for providing IT and network support (ensuring that all presenters have access to and can share any digital materials that are part of their presentation).
- Each team has received sufficient input to produce a concept of information, including relevant information about the context in which the scenario operation will take place and the planned operations themselves.

- Each team has been given a list of capabilities that are considered organic to their side's forces and that they can employ as part of their concept of information.
- Each team has been provided with a planning space: separate physical or virtual spaces, each capable of accommodating all 6–12 members of each team plus 2–3 members of EXCON. Ideally, these spaces will include projection and network capability so that EXCON can display slides and players can collaborate, store material on a shared drive, and share and discuss slides and other content. It is also preferable that each of the two spaces also be the workspace for the respective teams.
- An engagement space (physical or virtual) has been prepared. This should be a space in which all players from both teams and all of EXCON can come together. It should not be either team's planning space or workspace, to avoid "home field advantage." The engagement space should have projection capability and network access so that presenting players can use slides if they wish.
- Game boards (either physical or digital or both) and appropriate tokens, markers, or other symbology have been made available to represent the current state of the scenario as understood by each side, as well as intended progress (both for maneuver and for information). As envisioned, the game requires four boards: one for each team planning space, one shared board for depicting actions and operational progress in the engagement space, and a ground truth/master board maintained in the EXCON control space to allow EXCON to track events.
- Dice are available, with a surface or box on or into which they can be rolled (or a virtual substitute; there are many dice-roller mobile apps available).
- Relevant forms, templates, and scoresheets have been printed in sufficient numbers to support both players and EXCON for all turns.
- A timer is available. Ideally, this would be a large countdown timer that displays time remaining to the overall large audience, so that all players and presenters can see how much time is remaining in a segment. Failing that, a stopwatch, kitchen timer, app, or other substitute can be sufficient but will require the timekeeper to display warning cards to indicate waning time remaining. If a large computer monitor or projector is available, a timer application could also be displayed that way, provided that it is visible to all participants.
- Should the wargame be scheduled to be played remotely rather than with all players and EXCON members together at the same site, additional materials and preparations may be required.

2. Detailed Procedures for the Conduct of the Five Steps of the IWX Wargame

This section provides detailed instructions, procedures, and rules for the five steps of this wargame.

2.1. Detailed Conduct of Step 1: Receive Scenario and Situation Update

2.1.1. Overview of Step 1

During Step 1, teams receive an update on the current state of the operation; how (if at all) it is deviating from the planned concept of maneuver for their side; an estimate of the time in the scenario that will be covered by the turn; the current locations of available information capabilities (and notification of any capabilities which have been exhausted or attrited); the status of GREEN; and any updates on injects or changing scenario conditions. The Step 1 overview summarizes the current state and lays out the intended maneuver progress for the turn for each side. It provides players with a snapshot of the current state of play and where the (notional) commander of their force anticipates progress in the next period of play. Step 1 takes place simultaneously in each team's planning space.

2.1.2. Inputs to Step 1

During Step 1, members of the EXCON narration and scenario team update players on changes to the operation and environment; on expected events in the next turn; on the relevant status of GREEN; on the current location and status of available information capabilities; and on timing information (both in the scenario and deadline/events in the game schedule) for the coming turn.

In the first turn of the game, Step 1 may be unnecessary: Players should know the background of the context and how the operation is slated to commence.

In subsequent turns, EXCON personnel providing updates as part of Step 1 can answer questions about outcomes of the previous turn as needed, although most such information will have been covered in Step 5 of the previous turn.

The following will be needed for Step 1, all of which must be provided by EXCON:

1. game boards (physical, digital, or both) and appropriate tokens, markers, or other symbology. These should depict:
   - a snapshot of the current state, as known by each side
   - the intended maneuver and other efforts planned for this turn or the intended end state for this turn, for each side
   - the positions of each of the information capabilities available to the side
   - major GREEN elements and expected GREEN activity
2. a briefing or notes for a verbal update for each side detailing:
   - a rough timeline of intended progress and an approximation of the amount of time in the scenario that the coming turn will represent
   - intended maneuver and other actions planned by the friendly side for the turn
   - intended end state for the friendly side for the turn
   - any deviations to the original friendly maneuver plan for the turn
   - most likely/most dangerous course of action information regarding the other side
   - available information capabilities and where they are located, especially any changes due to capability exhaustion or attrition (see Section 2.5.4.4)
   - relevant expected GREEN actions or status
   - any injects that will occur during the turn (these should take the form of inject cards; see Sections 2.6.1.2 and 3.4.6)
   - deadlines and expected event times (where the players need to be and when) for the remaining steps in the turn.

Ideally, all of these materials will be preplanned/prepackaged for several possible storylines before the game and will only need to be adjusted based on variation in operational outcomes as necessitated by extreme success or failure by one of the teams of players or due to optional injects.

NOTE: Changes from previous turns are particularly important to note: changes in the maneuver plan, changes in available information capabilities (particularly changes due to capability exhaustion or attrition; see Section 2.5.4.4), changes in timespan represented by the turn, etc.

NOTE: Players can get hung up on the exact length of time represented by the turn and when in the turn their actions will take place. Try to keep some flexibility about how time is represented (next day or two, next few days, next week or so, next couple of weeks, etc.) and focus on the commander's intent, especially any objectives or decision points. Players should be encouraged to launch their actions from the presented baseline, but if the timing of their actions is contingent on progress of maneuver forces, actions should be tied to progress to or toward specific objectives rather than tied to specific points in time. This will make narration easier and prevent players from claiming that actions started prior to actions that have already been presented (by rule, actions begin in the order in which they are presented, though their durations may overlap).

Game boards should be managed by a game board keeper from EXCON (if one has been designated) or another member of the narration and scenario team.

OPTIONAL: If circumstances require that fewer than the default three actions per team per turn will be allowed in the turn, this should be announced during Step 1.

2.1.3. Player Activities During Step 1

In this step, players receive updates and information and take notes on new information as needed. Players ask clarifying questions about the updates they receive.

2.1.4. EXCON Activities During Step 1

EXCON provides the updated game board and the briefing or verbal presentation of updates, changes, and injects (injects are described in Sections 2.6.1.2 and 3.4.6). The goal is to provide a snapshot of the state of play as of the beginning of the turn (the conditions from which players should plan their actions), as well as what progress the commander expects the friendly force to make during the upcoming turn (the activities and objectives that the team's information activities should hasten, support, enable, or accomplish).

EXCON provides these updates to both teams separately but simultaneously, so the absolute minimum required EXCON personnel for the step is two. Ideally, two or (better still) three EXCON personnel drawn from the narration and scenario team should be present for the update to each team so as to have more expertise available to answer any questions that emerge and to take notes about questions (and the answers given). The total recommended EXCON personnel requirement for the step is between four and six.

NOTE: EXCON should not answer some types of questions that players ask, and should be careful in giving players new information. Questions about game procedures should be answered fully. But requests for information (RFIs) about past or planned future events should be answered only to the extent that an information OPT would have access to such information as part of situational awareness and other intelligence flows within that headquarters within the scenario. EXCON should also be careful with the invention of details that players should have access to but that are not in the materials provided, as these inventions must then be incorporated into the ongoing wargame scenario if they prove to be important details.
Any RFIs answered by any EXCON representative during Step 1 or at any other time should be recorded on an RFI matrix (notes indicating who asked what, when, and what they were told) and reviewed by the entire narration and scenario team so that any changes in the scenario world implied by the answer can be recorded and carried forward throughout the game.

OPTIONAL: The S3 role-player may attend the update and may present a portion of the update.

OPTIONAL: The GREEN portion of the update brief can be presented by a separate EXCON member responsible for GREEN, with that individual visiting both team rooms.

2.1.5. Outputs from Step 1

Step 1 should produce (1) the notes the players take to support their efforts in later steps and (2) the notes EXCON takes based on any exchanges of information with the teams.

2.1.6. Time Allowed for Step 1

Step 1 should be completed in 15–30 minutes or less. Teams can proceed immediately to Step 2, so any extra time allocated to Step 1 can be put to good use. Step 1 may be unnecessary for the first turn of the game.

2.2. Detailed Conduct of Step 2: Prepare to Present

2.2.1. Overview of Step 2

In Step 2, teams revisit their plans as needed and prepare their actions for the turn. This step takes place in each team's planning space. Teams compose an oral presentation for the S3 role-player for their side (for Step 3) and also marshal their arguments supporting these actions for engagement (Step 4). The presenting player for each action rehearses their presentation and makes sure they are ready to offer compelling explanations and arguments. This is also the step in which a team should identify non-organic assets, or unique permissions or authorities for special capabilities that their actions might require and will require additional approval in Step 3 from their S3 role-player.

2.2.2. Inputs to Step 2

At the outset of Step 2, players should have all the inputs they need. These include their existing plans; their understanding of the capabilities available to them or that they can request; all the scenario background materials; and their understanding of the current situation and the desired progress of the operation during the turn, the status of relevant GREEN groups, and the locations and status of available information capabilities, all based on their notes. Other possible inputs might include answers to any questions that the players come up with as they prepare, limited as above to answers to which they should be entitled (game procedural answers, or answers that an OPT in the HQ they represent would be able to get within the confines of the time represented by the turn).

2.2.3. Player Activities During Step 2

During Step 2, players update their overall plan (their concept of information) as needed in response to changes in the anticipated flow of the operation, unanticipated actions by the opposed side or team, and other changes in the scenario context. Players also prepare their actions for the turn. If an EXCON member is present, teams may ask questions or refer RFIs through that individual. If no EXCON representative is in the team room or if they do not know the answer to a posed question, teams may send their team leader to the EXCON control room to ask questions as needed. Teams should identify any non-organic capabilities or special authorities or permissions that will be required for their actions and cue EXCON to their intent to request additional capabilities or permissions. These requests will be adjudicated in Step 3 (see Section 2.3.4.2), but might optionally be adjudicated during Step 2, if Step 3 is omitted or folded into Step 2 (see optional rules in Sections 2.3.7 and 2.3.8).

2.2.3.1. Preparing Actions to Present

2.2.3.1.1. How Many Actions to Prepare

Teams may present no more than three actions per turn during Step 4. Teams may wish to prepare more than three actions in case one or more actions are not approved. Players may also wish to prepare alternative versions of actions (in case requested non-organic capabilities are denied), or prepare branches and sequels for their actions in case early actions do not succeed and set conditions for planned implementation of later actions. Players might also plan ahead and begin to prepare actions for later turns. Only actions approved in Step 3 may be taken in Step 4.

OPTIONAL: Players may be explicitly instructed to prepare more than three actions so that some can be downselected as part of the approval process in Step 3.

Actions begun in a previous turn and continued in a later turn must be presented again for approval in Step 3 and presented for adjudication during Step 4 in that later turn. Action previously attempted that failed may also be re-presented for approval and re-attempted.

OPTIONAL: Continuing actions may be considered as not counting against the action limit.

2.2.3.1.2. Rotation of the Presenting Player

Each team action is presented by a single player. That player is the presenting player for that action and must be the one who presents the proposed action in Step 3. If the action is approved, that player must be the one who presents the approved action for matrix discussion and adjudication in Step 4. If the action continues into later turns as a continuing action, that same player must continue to present the action.

Players must rotate presentation of actions through the entire team to ensure that each player has the opportunity to present, and so that exercise staff have an opportunity to observe each player's demonstrated learning. A player who has already served as the presenting player for an action may not be the presenting player for another action until all other players on the team have served as the presenting player. This rotation continues across turns and phases.

2.2.3.1.3. Materials Needed to Present a Candidate Action

The presenting player must be prepared to verbally explain and justify a proposed action to the S3 role-player in Step 3. To present a candidate action, players should have prepared remarks. If they wish, players may enhance their presentation with slides or support it with tokens or other representations to be placed on the game board (or moved on a digital game board). Action proposals must detail the following:

- target, TA or, audience, and location
- effect desired, including the location, time, and duration of the effect
- purpose of the action and how that purpose contributes to the commander's intent and the concept of maneuver for the phase
- forces or capabilities that will conduct or contribute to the action, where they will be located (on the game board), how they will get to that location, and when their activities will occur
- the desired end state, including what failure will look like and what success will look like
- an assessment plan, including MOPs and measures of effectiveness (MOEs), how those measures will be collected or observed, and how long after the activities/effects occur measures will be reported
- a list of at least three reasons the action will succeed (only three reasons can be presented in Step 4, but more might be presented in Step 3).

The presenting player should rehearse their presentation during Step 2, time permitting, so that other team members can provide them with feedback and to maximize their preparedness for Step 3.

OPTIONAL: EXCON (likely through the S3 role-player) may also ask players to prepare reasons the action might fail to execute or fail to have the intended effects. This might just be an element of the action approval package/presentation, or it might be a separate written task. **Note that, if used, this optional element should *not* be presented during engagement and matrix debate in Step 4**—the goal is not to make it easier for the other team to make rebuttals! Rather, the goal is to anticipate rebuttals and to promote critical thinking about ways in which the action might be approved, or to get to more realistic expectations about the action's likely success. If used, this optional element could also support the optional rule for self-critique described in Section 2.4.3.9.4.

2.2.3.1.4. Order of Presentation

Players should plan to present their actions in the order in which those actions are scheduled to begin, with continuing or ongoing actions presented first (see Section 2.2.3.1.5). Players should present actions in the same order in both Step 3 and Step 4 unless instructed to do otherwise. See more on order of presentation in Section 2.4.3.2.

2.2.3.1.5. Repeated, Recurring, or Ongoing Actions

Players may wish to conduct actions from previous turns again in later turns, either because the action is continuing or because the action previously fell short and they want to try it again. If a team wishes to repeat or continue an action, they need to present it again for approval in Step 3 and present it as part of engagement in Step 4. If little has changed, then repeat presentations may be somewhat abbreviated. However, if something significant about the environment or execution is different, the team should make a full presentation of the action.

As noted, a player who is designated as the presenting player for an action remains the presenting player for that action, no matter how many times the same action is presented.

2.2.3.1.6. Re-Presenting a Previously Disapproved Action

For actions that the S3 role-player disapproved in a previous turn but recommended for further development, the presenting player remains the same; that is, if an action is not approved during Step 3 in one turn and is presented for approval again in a subsequent turn, it must be presented by the same presenting player.

2.2.4. EXCON Activities During Step 2

EXCON engagement with the players is minimal in Step 2, as the step is primarily a planning, preparation, and rehearsal period for the teams. EXCON might have four roles in this step: (1) being available to answer questions that come up, (2) providing mentorship, instruction, or advice to players, (3) observing the teams to see what they are going to be proposing as actions, and (4) using those observations of planned actions to prep the S3 role-player, to begin thinking about how such actions might be scored on scoresheets during Step 4, and to begin planning for how such actions might affect the flow of the operation and the storyline (for Step 5 of the current turn and Step 1 of the next turn).

Step 2 requires a minimum of one EXCON person available to answer questions located in the EXCON control room. It is preferable to have at least two additional persons available to take notes and to observe the teams. EXCON personnel required for the step range from one to three, with three being ideal. The S3 role-player **should not** be one of the EXCON personnel remaining in the team room to advise, as doing so could compromise their credibility in the role of hard-nosed S3 and their approval or disapproval of actions in Step 3.

NOTE: Any RFIs answered by any EXCON representative during Step 2 or at any other time should be recorded on an RFI matrix and submitted to the narration and scenario team so that any changes in the scenario world implied by the answer can be recorded and carried forward throughout the game.

OPTIONAL: Instead of seeking to maintain the hard-nosed credibility of the S3 as a challenge or barrier in the process, the game could instead employ the S3 role-player as a mentor, offering advice to the team and trying to shepherd their actions to success. In this case, extra care must be taken in Step 3 to ensure that substandard actions are not approved.

OPTIONAL: If the optional rule omitting Step 3 is in use (see Section 2.3.7), then EXCON will need to make personnel available to adjudicate the availability of permissions and/or capabilities from higher echelons (as described in 2.3.4.2) during Step 2.

2.2.5. Outputs from Step 2

Step 2 should produce a set of candidate actions ready to be presented to the S3 role-player in Step 3. To present an action for approval, players should prepare notes for each oral presentation, as well as a supporting briefing or plan to use markers, tokens, or symbols on the game board as needed.

EXCON should gather these materials as outputs for use in preparing for steps 4 and 5. This might involve EXCON members (the S3s and/or the judges) viewing materials in a team's folders on a shared drive, or requesting print copies. EXCON outputs include a prepared S3 role-player who is aware of and ready for the actions that will be presented based on information passed from the EXCON notetaker or other EXCON observers or viewed on shared drives.

2.2.5.1. Battle Drills and Standing Operating Procedures

An additional possible output from Step 2 is one or more possible battle drills or standing operating procedures: things the side will do if and when certain conditions (usually bad conditions) come to pass. Such battle drills probably should be part of the larger plan prepared during the initial planning portion of IWX. Although such battle drills are likely not formal "actions" as understood in this wargame, they might be useful as arguments in the rebuttal or counterargument stage of Step 4. "Counteractions" and other unplanned responses to the actions of the other team are explicitly forbidden from being used in rebuttal or counterargument unless such actions are part of a battle drill that is documented in the team's plan and can be verified by the role-played S3. Having such battle drills or standing operating procedures (and sharing them with the S3) is not required, but they certainly could prove useful in the game (and in actual operations).

2.2.6. Time Allowed for Step 2

Step 2 should require 60–120 minutes to complete. If events are unfolding in a way that is consistent with a team's plans, then it should not take the team long at all to identify elements of their plan to propose as actions and prepare presentations for. However, if results have been going against a team or opposed play has caused events to turn in unanticipated directions, players may be scrambling to adjust their plan (and the corresponding game actions) to compensate. Recommended allocation of time for this step is 90 minutes, either to allow players some slack time or to allow them sufficient time to adjust and recalibrate. If later turns represent reduced amounts of time in the scenario, reducing the allowed preparation time to 60 minutes might make sense.

NOTE: Players can and should work on elements of Step 2 (and future Step 2s) during slack time, including excess time in other steps, time between steps, time between turns, and time before or after the formal exercise time for each day is complete.

2.3. Detailed Conduct of Step 3: Present Actions for Approval

2.3.1. Overview of Step 3

Step 3 takes place in each team's planning space. For each proposed action, the presenting player (and only the presenting player) briefs the S3 role-player on the audience and intended effects, capabilities involved, and details of (planned) execution. A different presenting player presents each candidate action. The S3 role-player approves or disapproves each action. Approved actions that require capabilities or permissions not organic to the team's side's forces will be immediately adjudicated as able to proceed or unable to proceed by EXCON based on a special outcome roll. On the S3 role-player's return to the control room, the S3 role-player back-briefs the narration and scenario team, as well as the judges, on the approved actions, and the judges review approved actions and begin to complete action scoresheets in preparation for Step 4.

2.3.2. Inputs to Step 3

Step 3 has numerous requirements from both players and EXCON. Teams must provide three or more candidate actions ready for presentation, each including all of the elements listed in Section 2.2.3.1.3, and each to be presented by a different presenting player. EXCON must provide a prepared S3 role-player who knows the current state of the operation and who is ready to be critical of the presented plans but is also prepared to make realistic approval or disapproval determinations.

The S3 role-player should have a number of blank proposed action approval checklist forms equal to the number of actions to be presented. Ideally, the S3 role-player will have some sense of the candidate actions they will be asked to approve, either from familiarity with the team's plan or from cueing from EXCON observers dedicated to that team. As discussed in Section 2.3.4.2, EXCON should be prepared to adjudicate any actions that cannot be approved by the S3 alone.

2.3.3. Player Activities During Step 3

Players present candidate actions to the S3 role-player for approval. Players present the actions in the order in which the actions would begin. Continuing or ongoing actions are presented first. Presentations should include all the elements listed in Section 2.2.3.1.3 and repeated below. Each action will be presented by the presenting player. Unlike the presentations in Step 4 (when *only* the presenting player may address the judges), in Step 3 the presenting player must take the primary role in the presentation, but other players on the team are allowed to help in answering any questions the S3 role-player may have.

If an action is a repeat action from a previous turn (either because the action is ongoing, the action was not previously approved, or the action was previously attempted but was unsuccessful), then the same presenting player who previously presented the action must present it.

Players will take notes on the feedback they receive from the S3 role-player that will help them refine their presentation of the approved actions in Step 4, or help them refine actions that were not approved should they wish to propose improved versions of those actions in later turns.

After receiving approval for their actions for the turn, players will make any final adjustments to the presentation of the actions (to include narrowing down the reasons that the action will succeed to exactly three reasons) for Step 4.

2.3.3.1. Materials Needed to Present a Candidate Action (compressed from Section 2.2.3.1.3)

To present a candidate action, the presenting player should have slides or notes for a verbal presentation supported by tokens or other representations to be placed on the game board that will allow them to detail the following:

- target, TA or audience, and location
- effect desired, including the location, time, and duration of the effect
- purpose of the action and how that purpose contributes to the commander's intent and the concept of maneuver for the phase
- forces or capabilities that will conduct or contribute to the action, where they will be located (on the game board), how they will get to that location, and when their activities will occur
- the desired end state, including what failure will look like and what success will look like
- an assessment plan, including MOPs and MOEs, how those measures will be collected or observed, and how long after the activities/effects occur measures will be reported
- a list of at least three reasons the action will succeed (only three reasons can be presented in Step 4, but more might be presented in Step 3).

OPTIONAL: As noted in Section 2.2.3.1.3, the S3 role-player may also ask players to prepare reasons the action might fail to execute or fail to have the intended effects. This might just be an element of the action approval package/presentation, or it might be a separate written task. Doing a self-critique like this has several possible benefits for players: First, it might help them to tweak their action and improve it to reduce its vulnerability to these specific criticisms; second, it might prepare players for counterargument if one of those criticisms is raised by their opponents; third, it promotes a critical mindset that will help them formulate their rebuttals of opposed actions during Step 4.

If used, this optional element should *not* be presented during adjudication of the engagement and matrix debate in Step 4—the goal is not to make it easier for the other team to make rebuttals! Rather, the goal is to anticipate rebuttals and to promote critical thinking about ways in which the action might be approved, or to get to more realistic expectations about the action's likely success. If used, this optional element could also support the optional rule for self-critique described in Section 2.4.3.9.4.

2.3.4. EXCON Activities During Step 3

Each S3 role-player will meet with their respective side's team and receive a briefing on their proposed actions. Each action will either be approved or disapproved. Approved actions can be unconditionally approved for presentation in Step 4, or conditionally approved pending certain revisions before Step 4. Disapproved actions will either be permanently

FIGURE 5

A Player Proposes an Action to the S3 Role-Player During IWX 20.2

Photo credit: Nate Rosenblatt, RAND.

rejected or conditionally rejected with advice to revise the proposal for a following turn.

OPTIONAL: Rather than approving or disapproving each action as it is presented, the S3 role-player may listen to all proposed actions and then approve their top three.

Each action must be presented by a different presenting player. A player may not repeat as the presenting player until all players have presented. The exception is for an action that was previously presented (either because it was disapproved, was approved and failed, or was approved and is continuing), in which case it must be presented by the same presenting player who previously presented it. If an action was previously approved or is continuing with details unchanged, the S3 may elect to ask the presenting player to review all the details as a full action proposal, may request a summary review, or may simply declare that the action continues to be approved.

NOTE: One player must be the presenting player and make the whole presentation, but the presenting player may confer with other team members in response to S3 role-player questions, especially when information capability expertise is distributed across the team.

Each S3 should approve no more than three actions. If the team has more than three actions, then the S3 may have to choose from the available actions (or force the players to choose). It is up to the S3 role-player to handle this. They should feel comfortable declaring that certain actions align better with command priorities as justification for approving them over other proposed actions. The S3 should also favor actions that reduce prospects for taking the story line out of bounds (see Section 2.3.4.4).

In addition to approving "good" actions and actions that do not threaten the storyline, S3s should also consider available resources. How much of each information capability is required to do everything the team is proposing? What was the status and location of those capabilities at the start of the turn? How much lift or other transportation is available? How much of an additional force protection burden does the action impose? These kinds of considerations can be used to force prioritization of actions.

The S3 role-player should make sure that the starting position of all information capabilities is properly noted on the game board and that presenting players refer to which capabilities will be used and how they will move from place to place as required. If the same asset needs to be in two places at the same time or an asset is required to travel an unrealistic distance in an impractical manner, then the S3 should use that to force the players to adjust their actions.

NOTE: The S3 should also confirm that the information capabilities being used have not had their availability terminated or reduced due to exhaustion or attrition; see Section 2.5.4.4. The S3 can also nominate capabilities that are being overused (or overrelied on) by the team to the narration and scenario team as candidates for exhaustion. Overreliance on a capability should not be grounds for disapproval of an action, but it should be captured and discussed with the narration and scenario and considered for removal from the availability list during Step 1 of the next turn.

The S3 should use *GA6: S3 Role-Player's Action Approval Checklist* (see Section 2.3.4.1) to help their evaluation of actions. All actions must meet a minimum standard of quality, even if this standard will keep a team from having three actions to present. **As S3, do not approve substandard actions!**

Step 3 requires an S3 role-player for each team (two total) and should include a notetaker/record keeper for each S3 (two total). Thus, this step requires between two and four EXCON personnel across the two team rooms.

OPTIONAL: Required personnel can be reduced if the S3 is responsible for taking their own notes, but they have a lot of responsibilities to keep track of in this step already.

2.3.4.1. S3 Role-Player's Action Approval Checklist

Approval of actions is a holistic process that can be based on the S3 role-player's expertise and judgment. However, to help structure that decision, and to help the S3 role-player provide constructive feedback likely to help players improve their later presentations, we provide a checklist in the downloadable game materials as *GA6: S3 Role-Player's Action Approval Checklist.* **We strongly encourage the S3 role-players to use the checklist.**

NOTE: A proposed action does not necessarily have to fully meet all checklist criteria in order for it to be approved; that is left to the S3's discretion. Review of the approval checklist is an opportunity for the S3 to serve as a forcing function for completion of all presentational elements. While an S3 may approve an action that does not include all elements, they should at least ask after missing elements and encourage the players to add or expand upon those elements. If an element isn't in the Step 3 presentation, it likely won't be in the Step 4 presentation unless the S3 takes steps to ensure that it is.

GA6: S3 Role-Player's Action Approval Checklist includes two elements for each criteria: (1) a check as to whether or not the elements were included in the presentation (for example, did the presenting player describe the TA and indicate where on the game board they are expected to be?) and (2) a go/no-go assessment of that element of the proposed action (for example, yes, there is a specified TA, but is it a reasonable audience to target?). The checklist includes the following items:

- Target, TA, or audience is described, including location; TA is not too broad or vague.
- Effect is described, including intended location, time, and duration.
- Purpose and connection to maneuver are described.
- Capabilities/forces that will contribute are described, along with where they need to be, when, and how they will get there; and these capabilities have not been exhausted or attrited.
- "Reality check": the proposed action is actually something that could be done by the acting capabilities.
- Desired end state is described, including a description of success and a description of failure.
- Assessment plan is described, including MOPs, MOEs, and sources.
- At least three reasons are offered for the success of the proposed action.

NOTE: Players often propose infeasibly broad TAs relative to the tactical influence capabilities available to them. The S3 can help reduce this tendency by being critical of overly broad TAs and demanding they be refined. Remind players that a TA is a person or group to be influenced, so the TA should either be (1) an individual who makes a decision they will personally act on (I will fight, I will retreat, I will surrender), (2) an individual who makes a decision that will determine others' actions (such as an officer or leader issuing orders), or (3) a relatively small and homogeneous group in a highly localized area.

2.3.4.2. Actions That Require Capabilities or Permissions That an S3 at This Echelon Could Not Approve

Some proposed actions may require capabilities outside the organic capabilities indicated as available to each team, and some proposed actions may require approvals or permissions at higher levels than the S3 at this echelon. If such approval or release of additional capability is pro forma or all but guaranteed, then the S3 role-player may approve the action, indicating that although they themselves cannot approve or release, it is sufficiently automatic that EXCON immediately makes a favorable adjudication. Similarly, if approval is completely out of the question, or if a real S3 wouldn't even give the OPT permission to ask, or if the approval timeline is longer than the time allowed within the turn, then the S3 can immediately reject the action.

If, however, there is genuine contingency and realistic uncertainty about whether certain capabilities will be made available from elsewhere in the force or permission given from a higher echelon, this can be resolved with an outcome roll during the presentation of actions to the S3. Players still present their action as if it would be approved; then, immediately following the S3's approval decision, an outcome roll is made to determine whether necessary capabilities and permissions are available.

To use an outcome roll to adjudicate availability or permission from higher up, first establish a target number, using Table 2 as a rough guideline. Basically, if availability or permission is likely but not a sure thing, then the target number is 7; if availability or permission is truly uncertain or contingent, along the lines of a coin toss, then the target number is 10. If availability or permission is unlikely but still possible, then the target number is 14. *GA6: S3 Role-Player's Action Approval Checklist* includes a worksheet for actions requiring additional capabilities or permissions.

NOTE: The "rule of three" also applies to non-organic capability requests. Teams can make no more than three requests for capabilities not on their force's organic capabilities list each turn. (Prior to the addition of the rule of three, some teams "fished" for special or additional capabilities with a flood of requests, hoping for at least one good roll.)

OPTIONAL: Rather than making the determination of the target number for an additional capability roll themselves, the S3 role-player may return to the control space and quickly consult with other members of the narration and scenario team to select a target number and then return to the team's space for a roll. This could be role-played as "let me confer with higher."

Table 2 provides some starting guidance for choosing a target number for outside capabilities or permissions. The target number could be further adjusted based on circumstances. The following are examples of possible adjustments that could be made to the target number by the adjudicating EXCON team/S3 role-player (circumstances that make approval more likely *lower* the target number; circumstances that make approval less likely *raise* the target number):

- OPTEMPO of operation generally slow (not much demand on other assets/HQs): –2
- Capability has been requested previously and not approved by a thin margin: –2
- Capability has been requested previously but not approved by a wide margin: +3
- Friendly force is under significant duress and supporting capabilities are all busy: +1
- Capability has been requested, approved, and used previously: either +2 (you've had your opportunity, don't overuse this capability) or –2 (capability has already been made available and so is still available).

TABLE 2

Example Starting Target Numbers for Availability of External Capabilities or Permissions

Requires out-of-echelon capabilities:	Starting Target Number
The desired capability is present in theater and is commonly but not always available	7
The desired capability is present in theater but is a somewhat busy capability and could easily be unavailable	10
The desired capability is low density/high demand and may need to come from outside the theater or is otherwise hard to get	14
Requires permission or approval from a higher level:	**Starting Target Number**
Permission or approval could come from only a few echelons up and the level of risk is low, but there is a chance the approving authority is too busy or will request more information	7
Permission or approval requires a long approval chain, with more voices that might not concur, or risk associated with the action is moderate and rejection based on risk is a real possibility	10
Permission or approval must come from the highest levels but is fairly routine; still possible that approval will take too long or that a high-level stakeholder will be risk averse	14
Risk associated with the action is relatively high, and approving authority may reject the action or seek more thorough risk analysis	14

NOTE: The way in which outcomes of previously proposed and rejected actions were described can dramatically affect conditional modifiers. For example, if an action was proposed in a previous turn and failed to meet a permission-based outcome roll by 1 or 2, and the explanation given was "approval came, but it was too late and too slow for you to conduct the action this turn," then on a later turn the roll could receive a big decrease to the target number, or the roll could be dispensed with entirely and the action could be simply approved for this turn.

If a proposed action requires both external capabilities and external permissions, then the S3 role-player can choose to resolve the uncertainty with either a single outcome roll or with two outcome rolls. Be advised that requiring two outcome rolls creates two possible points of failure and thus lowers the overall prospects for success for the action, but two rolls might be appropriate for an action with "easy" availability and "easy" permission. If both external permission and capability are required, perhaps use the higher of the two target numbers. This could be justified: If approval comes from a high level for this action, the capability will surely be made available. This would be an S3 decision and at their discretion.

NOTE: The S3 (in consultation with the narration and scenario team) should narrate why a request has been denied, and this piece of narration should be reported to and tracked by the narration and scenario team. The reason a request is denied may have impact on the availability of a capability in future turns. If a capability that has a high target number for availability is repeatedly requested and the request is denied, the S3 can provide feedback or narration that amounts to "stop asking, the answer is always going to be no."

While the guidelines laid out in this subsection are intended to address permission or capabilities to execute information activities required from outside the purview of the S3, they could also be extended to cover intelligence-collection assets or other requirements for observing and monitoring the outcomes of intended actions. Should the assessment plan for an action require assets or collections from organizations or forces outside the control of the current echelon, a similar process could be used, with an outcome roll determining availability. Alternatively, availability/assignment of intelligence assets could be included in a single roll, with either or both capability availability and higher-level approval.

NOTE: Failure to secure assets to monitor or assess an action may not prevent it from being approved; an action could still be approved and attempted with insufficient assessment at the discretion of the S3. This may result in reduced information about the outcome of the action during Step 4 (see Section 2.4.3.9.5).

2.3.4.3. Final Determination of Approval Status of All Actions

After presentation, discussion, and (if necessary) an outside approval or availability roll, the S3 role-player should provide a clear determination of the approval status of each action to the players. There are four possibilities:

1. The action is approved as is.
2. The action is approved with adjustment before Step 4. (In this case, the S3 needs to clearly state what adjustment is required, and whether the S3 needs to see confirmation of the adjustment before Step 4 begins.)
3. The action is not approved in this turn, but should continue to be refined and should be re-presented in a future turn.
4. The action is not approved in this turn (back to the drawing board).

OPTIONAL: Once both teams have had their actions approved, they share a preview of their actions with the other team. This preview does not include the three reasons they believe their action will succeed, but does include the same action template or approval slide that they brief to the S3. The purpose of this exchange of actions is to allow opposed teams more time to think critically about the action and to offer better rebuttals during Step 4. This is not intended to allow teams to change their actions in response to what the other team is doing. If that appears to be happening, EXCON should take steps to stop it, either explicitly forbidding changes to actions after the action exchange, changing the timing of the action exchange to be later in the overall workflow, or ending use of this optional rule.

2.3.4.4. Challenges Facing the S3 Role-Player

Playtesting and experience with the game have revealed that the S3 role-player is one of the most challenging, and one of the most important, roles in the game. An S3 role-player can come under pressures of different sorts depending on the scenario, the specific team of players they are working with, and their own personality and proclivities. Here are some of the challenges that S3 role-players for the game have faced, and the sorts of problems these challenges can lead to for the game if inadequately addressed:

- *Losing objectivity.* If the S3 becomes too much a part of the team they are role-playing for, this can lead to unfair bias on their behalf or providing advice and guidance that crosses the line from mentorship to actually taking over and suggesting or shaping the team's actions. This can reduce the quality of the learning experience for the players and give their team an unfair advantage.
- *Failure to gatekeep.* If the S3 is too lenient, then poor or unrealistic actions can be presented (and adjudicated) in Step 4. Weak actions in Step 4 can be frustrating for the judges, can result in uncomfortable interrogations of the presenting player by the judges, and can produce very high target numbers and thereby risk a lucky throw of the dice achieving a very high target number and rewarding players with success for a poorly conceived action. Such a situation can increase pressure on the narrator or push the storyline out of bounds.
- *Too agreeable.* If the S3 is too agreeable, this can lead to failure to gatekeep and the approval of substandard actions, as noted above; it can also lead to granting permission for outside capabilities that should have been the subject of rolls (see Section 2.3.4.2), or to agreement about what is possible, how a capability works, or changes to the concept of maneuver for the side that are inconsistent with the reality of the scenario or for the narration and scenario team's expectations. This can, in turn, lead to discon-

nects between what the players understand and what the narration and scenario team understands, which can threaten the storyline.

- *Insufficiently connected to the narration and scenario team.* If the S3 isn't part of or at least in regular contact with the narration and scenario team, the S3 can end up making decisions for the scenario team that then end up sticking and being problematic. Or, a disconnected S3 can make a decision that is reasonable but fail to report it to the scenario team, so that assumptions between EXCON and some of the players diverge, creating unproductive disagreement and distraction during Step 4 and making EXCON look less professional in general.

OPTIONAL: These challenges might be avoided in a variety of ways. Here are several possible approaches to preventing the emergence of these challenges or reducing their impact:

- *S3 role-player diligence.* An S3 who is intimately familiar with the scenario, is familiar with the functioning of a wide range of information capabilities, works to preserve their role as hard-nosed S3, and is diligent about informing the narration and scenario team about the things they've told the players (or, better, is diligent about asking the narration and scenario team before sharing facts with the players) will avoid most of these problems.
- *Swap S3s.* If there are concerns about the S3s losing objectivity and providing undue advantage to their team, rotate which role-player plays the S3 for which side. A simple rotation, swapping the role-players every other turn, would work fine for this. Such a rotation does risk continuity (for example, an action that was close to being approved by one role-player one turn for some specific reason may have that reason lost during a swap), but it reduces any benefits from difference between the role-players or differential treatment by the role-players.
- *Use a placard or name tag to distinguish role-player roles.* If the same individual is serving both as a senior mentor and as the S3, find some way to make clear (and to remind them and the players) which "hat" they are wearing. This could be a physical hat, or a placard or name tag, that they wear only when acting as the S3, which they remove or reverse when acting in a different capacity. This could help remind the role-player when they are acting as a mentor or pal, and when they are supposed to be hard-nosed and doing some gatekeeping.
- *Use an approval panel or board.* Instead of having the S3s be hard-nosed, make them senior mentors and allow them to advise their teams and assist them to a certain extent. However, then conduct Step 3 approvals with an EXCON panel of three approvers, ideally drawn from the narration and scenario team, with majority approval required for an action to advance to Step 4.
- *Assign a narration and scenario team "buddy" to the S3 role-player.* Whenever the S3 role-player goes into the team planning room, they should be shadowed by this buddy. Most of the time, the buddy would just observe and record the interactions between the team and the S3 so anything relevant can be reported back to other narration and scenario team members, but the buddy is also there for consultation, or to intervene immediately if the S3 says something that strays from the scenario reality or the intended storyline.

2.3.4.5. "Backroom" EXCON Activities During Step 3

The main EXCON activity in this step is completed by the S3 role-players assigned to each team. However, the rest of EXCON should not be idle. Using observations made by the EXCON notetaker for each team in Step 2 (if assigned), and **with input from the S3 role-players after they return from receiving action briefings, each EXCON judge should begin to fill out a copy of *GA8: Engagement Scoresheet*** for each approved action so that they are as complete as can be heading into Step 4. Specifically, in Step 3, judges should be able to make preliminary judgments about how difficult the action will be to execute and how likely it is to be effective, filling the parts of the scoresheet labeled 1, A, and B. In Step 4, judges can update these parts of the scoresheets as needed based on the actual presentation, rebuttal, and any other elements of matrix discussion (see Section 2.4.3.7).

The narration and scenario team should begin to prepare, for each action, a copy of ***GA7: Narrator's Preparation Worksheet*** (see Sections 2.4.3.8.2.2 and

2.4.3.9.2) based on the preliminary reports of the S3s.

The EXCON keeper(s) of the game boards should also ensure that sufficient tokens to represent possible effects of the various proposed actions are available in the engagement room. Game board keeper(s) should also prepare a hybrid representation of both RED and BLUE situational awareness to display during engagement. This depiction should avoid giving away unknown maneuver positions but should provide enough of RED and BLUE positions (possibly ambiguously) to allow teams to describe the locations of their actions.

2.3.5. Outputs from Step 3

Step 3 should produce no more than three approved actions for each team. Step 3 may also produce one or more candidate actions that are not yet approved but could be approved in future turns for the team to continue to refine. The approval status of all actions should be clear to the players (as per Section 2.3.4.3).

NOTE: For actions that are not approved but are recommended for further development, the presenting player remains the same; that is, if an action is not approved during Step 3 in one turn and is presented for approval again in a subsequent turn, it must be presented by the same presenting player.

2.3.6. Time Allowed for Step 3

Step three has two substeps: (1) the presentation of actions to the role-played S3 for approval and (2) final adjustments and preparations to present those plans in Step 4. Total allocation of time for both activities should be 60 minutes. The players can have whatever time is left after their actions are approved (or denied) by the S3 to finalize the presentation of those actions for the next step. The S3 and the rest of EXCON can use this residual time to prepare for a smooth and efficient Step 4. Under some optional rules (see Sections 2.3.7 and 2.3.8), Step 3 is omitted or conducted in stride with Step 2.

2.3.7. OPTIONAL: Omit Step 3

Under certain circumstances, almost all of Step 3 might be omitted. This might be necessary if adequate S3 role-players are unavailable, or it might be attractive to save time by omitting Step 3 and treating the presentation in Step 4 as both the practice of justification of proposed action to staff seniors and as engagement for wargame adjudication.

If Step 3 is omitted, then the process for ascertaining whether approval or capabilities are received from higher echelons (as described in Section 2.3.4.2) should take place during Step 2, with team leaders requesting an EXCON adjudicator to support such determinations as needed.

If Step 3 is omitted, information from Step 3 that informs Step 4—such as information about the pending actions by each team necessary for EXCON judges to prepare draft scoresheets (Sections 2.4.2 and 2.4.3.7) and for the EXCON narrator to prepare draft narration of outcomes (Sections 2.4.3.8.2.2 and 2.4.3.9.2)—will need to be provided by some other means. This could be accomplished by requiring team leaders to submit single-page summaries of their planned actions by the end of Step 2, or by having someone from EXCON (a notetaker, or someone acting as a mentor in the team rooms) report this information back to EXCON central (the judges and the narrator at minimum) by the end of Step 2.

2.3.8. OPTIONAL: Complete Step 3 in Stride with Step 2

Similar to the 2.3.7 optional rule to omit Step 3, Step 3 may also be folded into Step 2. Under this approach, the S3 meets with the team early in Step 2 and gets a preliminary summary of the proposed actions, quickly asking refining questions, making rolls to determine the availability of required external capabilities or permissions, and approving three actions. The team then further refines the approved actions in preparation for Step 4.

As a further option under Section 2.3.8, when presented with the team's candidate actions early in Step 2, the S3 can immediately approve one action as the best and top-priority action, but return to the control space with the remaining candidate actions for a consultation with the narration and scenario team. This allows for a quick EXCON review and assessment of the plausibility of the proposed actions and their possible impact on the scenario storyline. Actions that are unrealistic or excessively disruptive can be avoided, and other alternative actions can be approved. A return to the control space relieves the S3 of the burden of making that call entirely by themselves. Once the remaining two actions are approved, the S3 returns to the team space and notifies the team of the determination.

2.3.9. OPTIONAL: Pre-Approval Confirmation Brief

One way to encourage a higher quality of actions is to ensure that they nest into a higher-quality overall plan. One way to do this is with a formal confirmation brief before play begins. Optionally, a single Turn 0, or pre-game, pre-approval/confirmation brief panel could make Step 3 unnecessary in all turns (that is, if the pre-game plan is approved, then turn-by-turn approval of actions is deemed unnecessary). Or a pre-approval/confirmation brief could just be in addition to the Step 3 approval of each individual action.

If used, EXCON should provide guidance or a template for the pre-approval/confirmation brief to the teams of players. At minimum, players should be informed of the goals of the brief. This confirmation brief panel event has several goals:

- to get the players to clearly articulate their concept of support; how it connects to theater, country, and campaign objectives; and how the concept will unfold during the game
- to secure any additional approval, permission, or commitment of support required to execute the concept
- for the SME panel to provide feedback or instruction related to the concept of support.

Players should be encouraged to not try to give a detailed description of all actions planned for all turns of the wargame, but instead to provide a broader concept of operations or concept of support briefing.

2.3.9.1. OPTIONAL: Confirmation Brief Before a Panel

A confirmation brief is also an opportunity to more formally represent higher echelons of command or outside stakeholders in the game. The panel can be assembled from available high-echelon and outside stakeholder representatives, and can optionally be filled out with EXCON role-players simulating important stakeholders not represented or available at the exercise.

The following guidance can be provided to panel participants and used to structure the pre-approval/confirmation brief.

The panel should aim to provide three things:

1. **Education**
   - Explain, from the perspective of the stakeholder or staff echelon represented, what things would have to happen during or before an actual operation of this kind to get the plan and planned actions socialized, approved, and authorized.
   - Describe processes and timelines.
2. **Feedback**
   - Give a review of the proposed plan and proposed action from the stakeholder perspective.
   - Offer suggestions for improvement.
   - Give an assessment of how likely such a plan would be to receive support from the represented stakeholder if proposed in the real world.
3. **Approval**
   - For game purposes, what in the plan is approved versus what needs revision or needs to be scrapped entirely?
   - The panel should have a "bias toward approval" for game purposes: Only actions that a stakeholder would unambiguously reject should not be approved (though actions that would face barriers and challenges should have that noted as part of feedback).
   - The panel should make clear what needs to be changed for marginal plan elements to be approved.

The workflow for the pre-approval/confirmation could proceed roughly as follows:

1. The team presents their plan (30–60 minutes).
2. The panel asks any questions of clarification (10–30 minutes).

3. Each stakeholder represented receives 5–10 minutes in which to provide education and feedback related to the plan, with questions from the players as needed (5–10 minutes per panel member).
4. The panel quickly confers regarding plan approval (with the possibility of either rejecting or requiring revision to some plan elements while still approving the overall plan) (5 minutes).
5. The chair of the panel conveys the results of the approval discussion to the players (5 minutes).

The above two lists are available as *GA9: Guidance for Pre-Approval Panelists.*

2.4. Detailed Conduct of Step 4: Engagement and Matrix Debate

2.4.1. Overview of Step 4

In Step 4, all players from both teams come together in the engagement space, and teams alternate presenting approved actions. The same presenting player who presented actions in Step 3 presents them in Step 4. During engagement, each presenting player also presents three reasons why they believe their action will be successful. The opposing team then has a few minutes to discuss and presents three reasons why they believe the action will be unsuccessful or less effective than the presenting player has indicated. The presenting team then has two minutes to prepare and an additional minute to present up to three counterarguments to the rebuttal. (This is a form of what is called *matrix adjudication* in the wargaming community.)

EXCON has an opportunity to ask clarifying questions, and then EXCON quickly confers and finalizes scoresheets, giving the presenting player a final target number. The presenting player then makes an outcome roll, which will produce a degree of success or failure for the action. The narrator then describes what happens as a result of the action. Teams alternate presenting actions, engaging in matrix discussion, receiving EXCON input and target numbers, and making outcome rolls until all actions have been adjudicated. All aspects of Step 4 are governed by the rule of three: Teams present no more than three actions, presenters offer no more than three reasons for success, rebuttals consist of no more than three reasons the action won't succeed, and rebuttals are followed by no more than three counterarguments. EXCON participation is also governed by the rule of three: Each judge asks no more than three questions, and (if included) a GREEN representative asks no more than three questions.

2.4.2. Inputs to Step 4

In Step 4, both teams and all of EXCON meet in the engagement space. Step 4 is where all the outputs of the previous steps come together. Teams will need presenting players prepared to present and defend their team's actions, possibly including briefing slides or a plan to display the action on the game board. EXCON will need a panel of judges prepared to score and adjudicate the actions (three judges, one of whom should be the EXCON lead).

Ideally, **EXCON will already know what actions each team will present**, from the combination of EXCON observers during Step 2 and the back-brief from the S3 role-players at the end of Step 3. Each EXCON judge should have, for each action, **a copy of *GA8: Engagement Scoresheet* that they began filling out in Step 3** (see Section 2.3.4.5). The EXCON narration and scenario team should also be present, and they should have **a partially completed *GA7: Narrator's Preparation Worksheet*** for each action with narrative outlines for various possible outcomes (see Sections 2.4.3.8.2.2 and 2.4.3.9.2).

EXCON (specifically the map keeper, or whichever member of the narration and scenario team has been assigned responsibility for maintaining the maps) should set the map in the engagement room as a critical input to Step 4. A member of EXCON should act as timekeeper.

NOTE: This step requires full participation by all players and EXCON. It is the most exciting step of the wargame, and, as such, everyone should be available to witness it.

With regard to manning requirements for EXCON, as described in Section 1.5, minimum recommended roles assigned are as follows:

- the three judges (one of whom should be EXCON lead)

- an S3 role-player for each team
- the narration and scenario team, which should consist of
  - the narrator (who will describe the outcome of adjudicated actions and build results into the scenario storyline)
  - a notetaker assigned to each team
  - someone to manage the game board
  - someone with responsibility for keeping track of the facts of the scenario (the "reality master")
- someone to serve as timekeeper, managing the stopwatch or timer to constrain times for presentation and rebuttal
- an action officer or stage manager, who has the critical role of making sure participants know when to arrive each day, when to reconvene for the next step in the turn, how long lunch breaks are intended to be, and otherwise make the trains run on time
- individuals with responsibility for providing IT and network support (ensuring that all presenters have access to and can share any digital materials that are part of their presentation).

Thus, there are a total of 13 EXCON roles, some of which could be filled by the same individual; for example, a notetaker could also serve as timekeeper, or the reality master could also manage the game board.

OPTIONAL: The EXCON panel can also include an EXCON representative explicitly focused on and responsible for the GREEN perspective.

2.4.3. Player and EXCON Activities During Step 4

Because this step involves an iterative sequence of activities that involve both teams and EXCON, these activities are described in a single sequence rather than separated into player activities and EXCON activities, as is the case in other sections of this ruleset.

Table 3 lists the activities in sequence and notes the ruleset subsection in which they are described.

The various presentations, rebuttals, and counterarguments of the teams are timed to ensure fairness and to keep things moving. As described in the subsections that follow, the presentation and argumentation of an action uses this sequence and these time constraints:

- Presenting team presents (5 minutes—described in Section 2.4.3.3)

TABLE 3
Step 4 Activities

Activity	Time Limit (minutes)	Section
1. Reorientation to the game board and situation		2.4.3.1
2. Determine the order of presentation		2.4.3.2
3. Presenting team presents	5	2.4.3.3
4. Non-presenting team rebuts	4 to prepare, 2 to speak	2.4.3.4
5. Presenting team offers counterarguments	2 to prepare, 1 to speak	2.4.3.5
6. EXCON interrogates, completes scoresheets, prepares for outcome roll		2.4.3.7
7. Outcome rolls, outcome determination		2.4.3.8, 2.4.3.9
8. Repeat activities as necessary		2.4.3.11
9. Record all results and prepare for Step 5		2.4.3.12

- Non-presenting team prepares rebuttal (4 minutes—described in Section 2.4.3.4)
- Non-presenting team rebuts (2 minutes—also described in Section 2.4.3.4)
- Presenting team prepares counterargument (2 minutes to prepare—described in Section 2.4.3.5)
- Presenting team counterargues (1 minute—also described in Section 2.4.3.5).

OPTIONAL: Timekeeping times and tracking may be adjusted based on the equipment available and in a manner that suits special circumstances (for example, a presentation by a non-native English speaker) and the preferences of EXCON lead. This may include a "soft limit," whereby players are told when time expires but are allowed to finish making their point.

GA3: Judge's Responsibilities summarizes the responsibilities of the EXCON judges in Step 4 (and in Step 3 in anticipation of Step 4).

2.4.3.1. Reorientation to the Game Board and Situation

Before beginning the cycle of action presentation and adjudication, EXCON should set the game board to display a hybrid of RED and BLUE situational awareness and set the stage by reorienting players to the conditions on the battlefield and in the environment at the beginning of the turn, as well as describing major maneuver objectives for each side. This is not intended to be as elaborate as the situation update presented in Step 1 but rather a simple review to remind all participants of the starting positions for the turn.

NOTE: The reorientation to the game board must be preceded by an intentional effort to set the map. This should be done by the map master and in concert with whomever will deliver the reorientation, so that what various tokens represent is clearly understood. Setting the map takes a nontrivial amount of time and should not be rushed or an afterthought.

OPTIONAL: The narrator, the reality master, someone else from the narration and scenario team, or the EXCON lead could provide this reorientation.

FIGURE 6

Players Survey the Game Board Prior to Engagement During IWX 23

Photo credit: Steven Webber, RAND.

2.4.3.2. Determine the Order of Presentation

Teams alternate presenting actions. If, for some reason, one team has a greater number of approved actions to present, that team will go first. Usually, teams have an equal number of actions approved, and which team goes first will be decided by a roll-off (that is, a representative from each team rolls the dice [ties are rerolled] and the team with the higher roll presents first).

Teams should present their actions in the order in which they are scheduled to begin and in the same order in which they were presented in Step 3. Continuing/ongoing actions should be presented first.

Actions are **not** assumed to be simultaneous; actions presented first are assumed to have begun first. Actions with lengthy durations are assumed to overlap other actions with lengthy durations.

NOTE: If, based on role-played S3 notes or other observation, EXCON knows that one team's actions could preempt, disrupt, or otherwise affect the outcome of one or more of the other team's actions, EXCON may impose an order or allow things to unfold based on the assigned alternation of turns. If potentially preemptive actions would logically take place earlier in a turn (they are faster, would begin concurrent with an early maneuver movement, etc.), then EXCON should arrange for that action to be presented first. If timing of a potentially preemptive action is undetermined relative to actions it might disrupt (that is, it isn't clear which action would realistically start first), EXCON may choose to allow the actions to unfold in the order determined by the turn sequence, with benefits accruing to whichever side based on events occurring in the order in which they are presented. Guiding principles to determining EXCON decisions here are realism and fairness. The possibility of the different sides' actions interacting is part of why action outcomes are not fully narrated until Step 5. See Sections 2.4.3.9.3 and 2.5.4.1.

2.4.3.3. Presenting Team Presents

Action presentations may take no more than 5 minutes and will be controlled by a timer; any material not presented within the time limit will be excluded from consideration.

OPTIONAL: Time limits may be waived or extended by the EXCON lead. This may include a "soft limit" whereby players are told when time expires but are allowed to finish their point.

Only the presenting player may present. That player may consult quietly with other members of their team during their 5-minute presentation period, and other team members may help with audiovisual display (advancing slides, handing around handouts, adjusting pieces or icons on the game board), but only the presenting player may speak to the judges about the action.

NOTE: Ideally, the timekeeper will have a countdown clock with a large visible display within line of sight of the players. Should this not be the case, the timekeeper will need to be more active in terms of letting presenters know when their time is drawing down. This can be difficult to do without interrupting the flow of play. Absent (or perhaps to supplement) a large and visible countdown clock, the timekeeper should have full sheets of paper, ideally colored yellow and then red, emblazoned with "2 minutes left" and "1 minute left!" which they can wave or display when the corresponding interval has elapsed.

To present an action, the presenting player should make a presentation that details the following:

- target, TA or audience, and location
- effect desired, including the location, time, and duration of the effect
- purpose of the action and how that purpose contributes to the commander's intent and the concept of maneuver for the phase
- forces or capabilities that will conduct or contribute to the action, where they will be located (on the game board), how they will get to that location, and when their activities will occur
- the desired end state, including what failure would look like and what success will look like
- three reasons the action will succeed.

NOTE: Some teams struggle to offer useful reasons that their action will succeed. Good advice for such teams is that they should consider aligning their reasons with the two sides of the scoresheet the judges will complete to score the action. The judges complete the "Difficulty of Execution" side of the scoresheet and the "Likely Effectiveness" side of the scoresheet. Players should consider making one or two of their reasons describe why the action will be performed successfully or why it is easy or at least straightforward to execute, and then one or two reasons explaining why the action will have the effects they intend (that is, why it will be effective). This advice can be given to the players during the in-stride AAR during Step 5 (see Section 2.5.4.3), or at other times by a senior mentor or the S3 role-player.

If time expires before the presentation is complete, the EXCON lead will consider whether there was an unavoidable interruption (such as an out-of-turn question by a distinguished visitor, or a delay due to technological problems). In the event of such an interruption, the EXCON lead may extend time as appropriate. If the EXCON lead deems that there was no such disruption, then the presentation ends immediately upon expiration of the timer.

During the presentation (and perhaps earlier, based on previews of actions from early intel on planned actions), the three EXCON judges should begin their scoresheet scoring. The scoresheets and scoring procedures are described in Section 2.4.3.7.

2.4.3.4. Non-Presenting Team Rebuts

When the presenting player concludes their presentation, the timer is immediately reset, and the opposing team has 4 minutes to prepare their rebuttal and an additional 2 minutes to offer it. The timekeeper will track this as two separate 4-minute and 2-minute periods, making a clear announcement of which period they are in. Teams do not need to wait for the initial 4-minute discussion period to be exhausted before offering their rebuttal and may proceed immediately to presentation of rebuttal when they are ready to do so, cueing the timekeeper to begin the countdown for their 2-minute presentation period. If the non-presenting team has not completed their rebuttal when the time allotted elapses (barring adjudication of interruption or special circumstances by the EXCON lead), they must immediately stop. Ideally, the game is played with a countdown clock visible to all players. If such a countdown clock is not available, the timekeeper should display a time warning card at 1 minute left.

OPTIONAL: Instead of two separate periods of 4 and 2 minutes, the non-presenting team may be given a single period of 6 minutes to discuss and present their rebuttals. This might be appropriate if the game is using a high-visibility countdown timer that shows the players their remaining time counting down.

To prepare the rebuttal, the non-presenting team may discuss among themselves, but only a single spokesperson may present the reasons—no more than three of them—that the action will fail or be less successful. These rebuttal reasons may take any form (a strong statement of fact drawn from the scenario materials, an assertion based on human psychology, a rhetorical question to the presenting team [which they are not allowed to answer until the counterargument]) and may address any reason that the action might be less successful (reasons related to information activity execution, design flaws in products, mistaken impressions about the TA, etc.).

However, **rebuttal reasons may not be actions**. This is a rebuttal, not a "counteraction," and the team must provide reasons the action will fail or be less successful because of the context, not because of something the rebutting team's forces are going to do in response. As explained in Section 2.2.5.1, the only exception to this rule is if the rebutting team has a battle drill or standing operating procedure in place (for which they can produce documentation and which their S3 will verify) that would lead to a response action. If a team has a relevant and approved battle drill or standing operating procedure, then one of their rebuttal reasons may make reference to efforts or effects that battle drill would produce.

NOTE: Frequently, teams and players struggle to come up with good rebuttals. Good advice for players is that, while any line of argumentation is open, when players in other IWXs have made good rebuttals, they have usually followed one of these five lines of reasoning:

1. The action will be harder to execute than the presenter suggests (and why).
2. The action will be less likely to succeed than the presenter suggests (and why).
3. The action may succeed, but it will have less of an effect than the presenter indicates (and why).
4. The action may succeed, but it won't matter; that is, the effect they are seeking from the target audience won't actually impact anything about the operation (and why).
5. Some of the assumptions made or facts claimed by the presenter are false (capabilities or forces aren't where they said they are, a capability they are employing doesn't work like they are saying it does, the target audience they are influencing isn't as amenable to their desired behavior as they suggest, processes described couldn't possibly be

completed in the time allowed); if you go this route, have evidence from the scenario briefings or RFIs, or from capability specialists.

If players are struggling to come up with counterarguments, they should consider quickly reviewing these five lines of argumentation to see whether and how one or more might be applied in their situation.

This advice can be given to the players during the in-stride AAR during Step 5 (see Section 2.5.4.3), or at other times by a senior mentor or the S3 role-player.

OPTIONAL: Another way to improve the quality of rebuttals is to use the optional rule first presented in Section 2.3.4.3: Once both teams have finalized their actions (late in Step 3), they exchange previews of their actions with the opposed team. This will give both teams more time to think about and craft possible lines of critique to employ in their rebuttals. Again, this optional rule is not intended to allow teams to change their actions in response to what the other team is doing.

OPTIONAL: If players are offering rebuttals that are wild, unrealistic, or out of compliance with offered guidance, the first time this happens, EXCON lead can interrupt the rebuttal and shut down the offered rebuttal, reminding players of the provided guidance and directing them to steer the rebuttal in the correct direction.

As the non-presenting team offers their three rebuttals, the EXCON judges may take notes on or further amend their scoresheets.

2.4.3.5. Presenting Team Offers Counterarguments

The non-presenting team does not get to have the final word. After the non-presenting team completes their rebuttals, the presenting team may quickly offer counterarguments. The presenting team will have 2 minutes to confer and prepare, and then a third minute in which the presenting player (only) will present counterarguments. The timekeeper will set the clock for 2 minutes as soon as EXCON lead announces the beginning of this substep.

FIGURE 7

Players Face Off During Step 4, Engagement, During IWX 23.3

Photo credit: Jonathan Welch, RAND.

OPTIONAL: The timekeeper both announces the beginning of the substep and starts the clock.

Counterarguments are limited in scope to being responses to the rebuttals from the non-presenting team. New points or issues may not be raised; only disagreements with or refutations of the rebuttal arguments are permitted. And, as is the case for the rebuttal arguments, counterarguments may not be actions. Only actions that are part of documented and approved battle drills or standing operating procedures may be presented as a counterargument. No more than three counterarguments may be offered. This could be one counterargument for each rebuttal reason, or three counterarguments directed at a single particularly telling or problematic rebuttal argument.

As the presenting team offers their final counterarguments, the EXCON judges continue to take notes and adjust their scoresheets as appropriate.

2.4.3.6. Secret Actions

This ruleset intentionally ignores the fact that some actions taken by one side or the other would be unknown to the opposed side and would remain so, even after adjudication. Such things might include deceptions, various aspects of operations security or signature management, or even very closely targeted influence efforts. These "secret" actions are revealed to all participants at the time they are presented, and their effects (which would presumably also remain unknown, at least under some conditions) are also revealed as part of adjudication.

When one team conducts an action whose result should remain unknown to the opposed side, all players should be informed of this by EXCON. Players should be advised not to act on information that they received during Step 4 or Step 5 but that they would not otherwise know. Players are prohibited from changing their actions to respond to or to counter actions by the other team that remain undisclosed *within the scenario*, even if they have been disclosed within the adjudication of the game. Players may need to be reminded (or re-informed) of this prohibition after the discussion and adjudication of any highly secretive action during Step 4. The reminder or admonition that "what happens in the engagement room stays in the engagement room" can be invoked to remind players not to act on knowledge about secret actions that they would not have in the scenario.

EXCON can enforce this separation between "player knowledge" (things players know from the open discussion of actions and results in matrix adjudication) and "scenario knowledge" (things that members of the information OPT or equivalent would know at this point in the scenario) in a number of ways. First, the maneuver aspects of the scenario follow and will remain on the paths laid out in the various established possible storylines, and these will not be changed by information that should not be known to those forces (for example, even if a deception is discussed during the engagement step, unless it is adjudicated to have been revealed or failed, the opposed side's maneuver forces will continue to act as if they believe the deception). Second, the S3 role-player can push back in Step 3 on any actions that rely on knowledge that should remain unknown and decline to approve these actions.

NOTE: How to treat secret actions was a matter of contention in initial game design discussions, but open discussion and display of secret actions has not caused any problems in any of the initial playtests or in several years of IWX play experience. This is likely for several reasons. First, players already understood that they were getting some information in the engagement room that they would not realistically have had in their staff position role. Second, the structure of adjudication, while allowing counterarguments, explicitly forbids counteractions, so there is no opportunity (and thus no temptation) to immediately take an action to counteract the secret action. Third, players understood that they represented a specific staff section, and that it would be the responsibility of the intelligence staff to uncover adversary deceptions and secrets and notify the commander and staff accordingly.

2.4.3.7. EXCON Interrogates (Optional GREEN Input), Completes Scoresheets, Prepares for Outcome Rolls

Following the presentation of the rebuttals, the EXCON judges finalize their scoresheets. Ideally, judges will have begun Step 4 with a partially and preliminarily completed scoresheet for each action based on reporting from other EXCON observers in Step 2 and Step 3 (especially from the S3 role-players) and so should be able to complete their scoring fairly quickly.

If the judges wish to do so, they may make any comments and ask any clarifying questions of the presenting player or the rebutting spokesperson; judges can ask questions of either side. Possible questions might include requests for clarification regarding possible unintended consequences.

NOTE: The focus should remain on the players and their actions. EXCON questions, either from the judges or GREEN, should be governed by the rule of three: no more than three questions from each EXCON panelists (of course, exceptions can be made for follow-up questions or instances where the player hasn't understood the question). Judges should ask questions intended to try to get players to provide more information or detail. Judges should not ask questions of the "stump the chump" variety or questions that are intended to show that the judge knows more than the presenter; once an ignorance of how capabilities actually work (for example) has been revealed, the judge may offer a comment about the way capabilities really function as a teachable moment, but should not hound the presenter by drilling down on a topic they are clearly not prepared to address.

OPTIONAL: At this time, an EXCON representative for GREEN may make relevant observations about likely GREEN response or may ask questions of the presenting player or rebuttal spokesperson. These observations might also include thoughts about possible unintended effects or tertiary effects. Again, questions should be governed by the rule of three and should not wax into "stump the chump."

Each judge completes their copy of *GA8: Engagement Scoresheet*, scoring the difficulty of executing the action on the left side (part 1 of the scoresheet) of the sheet and scoring the likely effectiveness of the action, were it to be successfully executed, on the right half of the sheet (parts A and B of the scoresheet). These scores will provide input for two different rolls: The "Difficulty of Executing" portion of the scoresheet seeks to create a composite measure of how difficult it would be to actually execute, or perform, the intended action, leading to MOPs. The "Likely Effectiveness" section assesses the prospects for the planned action, if successfully executed, to actually achieve the desired effects (so, the MOEs). The two scores can vary independently of one another. An action may be easy to execute but unlikely to have much effect, or an action could be complicated to execute but very likely be effective if successfully completed. On both parts of the scoresheet, low numbers are preferred by the action team: Low numbers in the "Difficulty of Performance" section represent low difficulty (the effect is easy), and low numbers in the "Likely Effectiveness" section represent low barriers to effectiveness (the action is likely to work).

In addition to scoring the difficulty of execution and the likely effectiveness for the action (in the areas labeled 1, A, and B on the scoresheet), on both sections of the sheet, judges score the strength of the presentation of the action—including the debate and discussion that followed the initial presentation of the action—to determine a debate modifier for each section. (See the parts of the scoresheet labeled 3 and D.) Judges add the debate modifier to the overall target number for each section. Poor performance in the presentation and debate will result in a modifier that increases the target number, and good performance will result in a modifier that decreases the target number. (Again, lower target numbers represent a greater chance of success on the outcome roll.)

NOTE: Bias toward success. When in doubt, judges should favor target numbers that are slightly lower, as this will create a bias toward success for actions. If players have planned well, they prefer to see their actions succeed, and learning is reinforced by success.

The scoresheets support the judges in target number determinations of the difficulty of execution and the likely effectiveness of that action if executed successfully, both modified as needed based on the discussion. The scoresheets are optional, and SME judges could instead make holistic determinations of the difficulty and related target numbers.

NOTE: The presentations are required to include an assessment plan, but the quality of the assessment plan is not one of the scoresheet criteria. This is because assessment has no bearing on the success or failure of the action! Assessment *does* bear on whether the players would know whether their action has succeeded or failed, and that is discussed in Section 2.4.3.9.5.

2.4.3.8. Adjudicating the Execution of the Action

After the judge's scoresheets are completed, the action is ready for adjudication. Adjudication will take place via two outcome rolls: one for the execution of the action and (if there is at least a partial execution) one for the effectiveness of the action. This section describes the process for adjudicating the execution of an action.

2.4.3.8.1. Judges Meet to Determine Outcome Roll Target Numbers

After the judges have completed their scoresheets, they should have each checked one box in the parts of the scoresheet labeled 1 (in the "Difficulty of Execution" section) and A and B (in the "Likely Effectiveness" section). Judges should add the total score for the boxes checked and reference that number in the table for that section (as instructed in parts 2 and C of the scoresheet). For example, if, after listening to a presentation, a judge decided that all six difficulty-of-execution criteria were of moderate difficulty, checking all the "moderate" boxes, that would add up to a total score of 12. The judge would then look up 12 in the table, finding that a total score of 12 produces an execution score of 10. This would then be modified by the judge's view of the presenter's performance in the discussion and debate, with discussion that made the action appear easier to execute reducing the score, and discussion that suggested the action would be harder to execute increasing the score. The difficulty-of-executing score plus the debate modifier produces that individual judge's target number for the execution outcome roll. Figure 8 depicts this example visually.

FIGURE 8

Example of Completed Judge's Scoresheet for Difficulty of Executing the Action

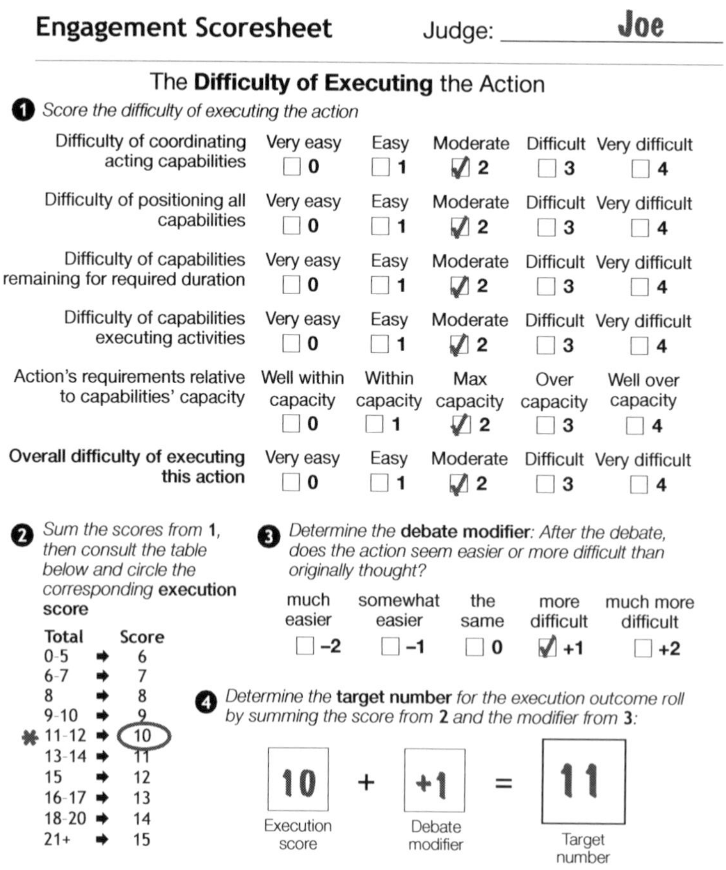

Engagement Scoresheet Judge: Joe

The **Difficulty of Executing** the Action

❶ *Score the difficulty of executing the action*

Difficulty of coordinating acting capabilities	Very easy ☐ 0	Easy ☐ 1	Moderate ☑ 2	Difficult ☐ 3	Very difficult ☐ 4
Difficulty of positioning all capabilities	Very easy ☐ 0	Easy ☐ 1	Moderate ☑ 2	Difficult ☐ 3	Very difficult ☐ 4
Difficulty of capabilities remaining for required duration	Very easy ☐ 0	Easy ☐ 1	Moderate ☑ 2	Difficult ☐ 3	Very difficult ☐ 4
Difficulty of capabilities executing activities	Very easy ☐ 0	Easy ☐ 1	Moderate ☑ 2	Difficult ☐ 3	Very difficult ☐ 4
Action's requirements relative to capabilities' capacity	Well within capacity ☐ 0	Within capacity ☐ 1	Max capacity ☑ 2	Over capacity ☐ 3	Well over capacity ☐ 4
Overall difficulty of executing this action	Very easy ☐ 0	Easy ☐ 1	Moderate ☑ 2	Difficult ☐ 3	Very difficult ☐ 4

❷ *Sum the scores from* **1**, *then consult the table below and circle the corresponding* **execution score**

Total		Score
0-5	➔	6
6-7	➔	7
8	➔	8
9-10	➔	9
* 11-12	➔	(10)
13-14	➔	11
15	➔	12
16-17	➔	13
18-20	➔	14
21+	➔	15

❸ *Determine the* **debate modifier**: *After the debate, does the action seem easier or more difficult than originally thought?*

much easier	somewhat easier	the same	more difficult	much more difficult
☐ −2	☐ −1	☐ 0	☑ +1	☐ +2

❹ *Determine the* **target number** *for the execution outcome roll by summing the score from* **2** *and the modifier from* **3**:

10 (Execution score) + +1 (Debate modifier) = 11 (Target number)

The judges then quickly confer regarding their scores and debate modifiers for the "Difficulty of Executing" side of the scoresheet. If they reach a consensus for these values, then the process moves forward with a consensus target number, as calculated in part 4 of the scoresheet. If they cannot reach consensus, then consensus is imposed by taking the median score and debate modifier (that is, the middle values among the three judges' scores and debate modifiers).

In addition to announcing the target number to the teams, both EXCON lead and the scenario and narration team (or associated notetakers) should record the target number for the execution outcome roll. A record of target numbers and outcome rolls can be useful when reconstructing events for narration, for after-action discussion, or for future calibration of scoresheets as part of continuous playtesting and game updating.

NOTE: When the judges meet to determine the target number for the execution outcome roll, they should *also* determine a preliminary target number for the effectiveness outcome roll, by comparing their individual scores and debate modifiers for effectiveness—parts C and D of the scoresheet—and either reaching consensus or using the middle values. The target number for effectiveness reached at this point is preliminary because it will be modified by the margin of success for the execution outcome roll, as discussed in the next section and in Section 2.4.3.9.2.

2.4.3.8.2. Margin of Success for the Execution Outcome Roll

Before the presenting player can make their outcome roll for the difficulty of executing of their action, the EXCON judges and narration and scenario team should have some understanding of what the roll might mean. In general, a roll the exceeds the target number indicates success, and a roll that falls short of the target number indicates failure. But what if the roll exceeds or falls short by a wide margin? by a narrow margin? hits the target number exactly? These are questions related to *margin of success*.

Table 4 depicts possible margins of success (or failure) for the execution roll and their interpretation. Each of the different margins of success has two impacts: The first is on how the execution of the action is narrated, and the second is on determining whether a roll for effectiveness will be allowed and what the target number for the roll will be. Margins of success range from a *remarkable success* at the top of the table to *failure* at the bottom. A remarkable success should be narrated as a nearly flawless execution under what turn out to be highly favorable circumstances, and the target number for the subsequent effectiveness roll should be reduced by 2. A failure should be narrated as performance so disrupted by circumstances, enemy action, or bad luck that the action effectively did not take place, and so no roll for effectiveness is allowed.

TABLE 4
Margins of Success for Execution and Their Interpretation

Margin of Success	Level of Success in Execution	Modifier to Target Number for Effectiveness Roll
Beat target number by		
6+	Remarkable success in execution	–2
2–5	Full success in execution	–1
0–1	Nearly successful execution	None
Fall short of target number by		
1–2	Near miss/insufficient effect	Not applicable
3+	Failure/no positive effect	Not applicable

OPTIONAL: EXCON may opt to contextualize the target number prior to the presenter actually rolling the dice, so as to remove some of the abstraction of the adjudication for players who are not as familiar with its mechanics. Judges might do this by making statements such as, for execution:

> We assess that your action is easy/medium/hard to accomplish, because of A, B, and C. Therefore, your roll is ____, which means a probability of ____ %.

And, for effectiveness:

> We assess the likelihood of you achieving your desired effects to be low/medium/high. With a roll of ____ on your first roll for performance, your target roll is now ____ for effectiveness.

OPTIONAL: The differences between the actual roll and the target number for different margins of success can be adjusted to reflect that an action is all or nothing, or is high-risk/high-reward, or for actions where partial success of execution is more likely, at the discretion of EXCON.

2.4.3.8.2.1. OPTIONAL: Critical Failures and Astounding Successes

Two optional rules are to have very poor rolls for either execution or effectiveness indicate extraordinary failures and to have very good rolls indicate extraordinary successes. The decision to include one does not necessitate the inclusion of the other. The rules for critical failures and astounding successes are as follows:

- Critical failures: An outcome roll total of 3 or 4 on the dice indicates a critical failure: The action's execution or effectiveness (as appropriate) has failed spectacularly. As an additional option, rolls that miss the target number by a wide margin, perhaps 7 or more, can also be considered critical failures. This possibility should be determined by EXCON prior to rolls. The EXCON narrator should describe not only a complete failure to execute or to achieve desired effects, but some sort of additional bad consequence. For example, for a critical failure in execution, perhaps the airlift moving the information capability to the action location is lost, creating a personnel recovery challenge and a reduction in the availability of that capability for the rest of the game. For an effectiveness critical failure, perhaps the action prompts some sort of backlash from the TA, enraging them when the intent was to pacify them. Or the additional consequence of a critical failure could have nothing to do with the execution or the likely effectiveness of the action but instead be a powerful exogenous effect (in game terms, a spontaneous inject), such as an atrocity on the part of the acting side being uncovered and publicized, or an important religious figure in another country making a huge speech that undermines the planned action. This can be especially attractive if a similar inject was planned for the next turn anyway—EXCON (the narrator) can uncork some planned misfortune and blame it on the dice.
- Astounding successes: An outcome roll total of 17 or 18—unless a 16, 17, or 18 was required for success at all—indicates an astounding success in execution or in effectiveness: Conditions favor execution or effectiveness to an extreme degree, and the action is executed brilliantly or is incredibly effective. As an additional option, rolls that beat the target number by a wide margin, perhaps 7 or more, can also be considered astounding successes. Because this outcome is a consequence of dice, the explanation of events leading to this outstanding success should include elements of good fortune: synergistic effects from chance environmental factors, unexpected but coincidentally beneficial behavior from an opposed side's subordinate commanders, or some similar unpredictable but beneficial result. For execution, the performance of the critically successful action can be described as going flawlessly and happening easily.

NOTE: For an outcome that has a target number of 16, 17, or 18, astounding success is not possible. An action with a target number of 16 or more is so difficult or unlikely to achieve its intended effects that any success is remarkable, and teams that are undertaking efforts with such limited prospects for success should not be rewarded further for luck.

NOTE: Again, the inclusion of both critical failures and astounding successes is optional, and the decision to include one does not necessitate the inclusion of the other. It would be perfectly reasonable, for example, to include critical failures but not allow astounding successes. This would be consistent with the pessimistic adage of Murphy's Law: What can go wrong will!

2.4.3.8.2.2. Preparing for Narration Based on Margin of Success

Playtesting and IWX experience have revealed that the narration of outcomes after rolls of the dice is extremely important, and that advance preparation of possible narration makes delivery of the actual narrative of action outcome smoother.

The narrator should prepare notes about how they might narrate each of the possible margin of success outcomes of a roll for each action for both execution and effectiveness. The narrator can't know in advance what the dice will deliver and so must be prepared to describe all possible outcomes, no matter how unlikely. See Section 2.5.4.1 for additional guidance on narrating outcome rolls. The point here is that the narrator should prepare notes to support different possible outcomes in advance and in relation to the various possible margins of success. A worksheet for making notes related to narrating different possible outcomes is provided in *GA7: Narrator's Preparation Worksheet*.

As part of this process, the narrator can also consider narrating exhaustion or attrition of information capabilities. Attrition of capabilities should be reserved from extreme failures of execution (including critical failures). When an outcome roll for execution misses the target number by a wide margin, one possible way to narrate the execution failure is to describe some kind of mishap that befell the required capability or damage that it sustained that prevented performance. If such a narration is made, a determination should be made in Step 5 if the damage is such that the capability is no longer available (see Section 2.5.4.4).

In contrast, the exhaustion of a capability could be narrated with either an execution success or an execution failure and should be driven by the level of demand placed on the capability rather than just an extreme dice roll. Exhaustion should only be narrated when such an outcome is plausible, and when the S3 or other member of EXCON has nominated the capability for exhaustion because of heavy use of the capability by the team. Capability exhaustion (temporary or permanent) is one way to herd teams away from a favorite capability or one that is becoming a crutch and toward something else (see Section 2.5.4.4). Take care not to narrate exhaustion of a capability that a team needs for an action later in the same turn until after that later action; the goal of capability exhaustion is to drive teams to plan a more diverse set of actions, not to deprive them of an already approved action.

2.4.3.8.3. Outcome Roll and Determination of Execution

With a target number determined and due consideration given to margin of success and possible narration, the EXCON lead announces the target number. The EXCON lead should provide more information than just a target number, also noting the judges' overall assessment of the difficulty of executing the action and the probability of beating the target number (see Table 10). Here is a sample statement:

> The judges have determined that the target number for execution of this action is [X]. This is because we deemed this action to be [very easy, fairly easy, moderately easy, moderately difficult, difficult, or very difficult] to execute. This corresponds to a [X] percent chance of success.

The presenting player now makes an outcome roll for the execution of their action, by rolling three dice and summing them, as described in Figure 3. This outcome roll total is then compared to the target number.[4]

The results of this outcome roll determine whether the action was successfully executed and to what degree (margin). The EXCON lead announces the margin of success for the execution of the action, and then the narrator offers a brief description of the execution. Unless the margin of success is "full success" or better, the narrator should offer some explanation for why performance fell short of a full success. Depending on the circumstances, this explanation might include environmental or contextual factors (e.g., strong winds swept leaflets off target), enemy action (e.g., radio transmitters were damaged in a kinetic strike and took too long to restore to functioning), bad luck (e.g., a vehicle crash prevented the executing capability from arriving at the target location in a timely fashion), or poor performance

[4] Section 2.4.3.8.3.1 presents optional rules related to rerolls. If a team has a reroll available and their initial roll falls short of the target number, they should announce their intention to use the reroll and reroll the dice.

by the executing forces (e.g., translation took longer than expected and the final product contained confusing translation errors).

NOTE: Narration for the execution roll should focus on execution and performance, consistent with MOPs. This narration is only of whether or not the executing forces successfully executed the action; the effects of the action and corresponding MOEs are determined by a second roll.

Notetakers should record the announced target number, the actual outcome roll, and the key elements of the description of the execution outcome.

Players should also take notes on when and why the execution of their action succeeded or failed. Was it because of bad dice rolls? Could they have improved their chances with better presentations or stronger rebuttals? These notes will help them adjust their plans in future turns of the game.

2.4.3.8.3.1. OPTIONAL: Rerolls

Any game involving dice includes an element of chance. Dice follow a probability distribution (shown in Table 10), but each individual roll of the dice is fully vulnerable to the winds of fortune. Luck represents uncertainty in the battlespace, be it fog, friction, chance, or Murphy's Law ("What can go wrong will"). However, consistent bad dice luck by one team can sour what would otherwise be an engaging and entertaining exercise experience. The option for rerolls mitigates slightly against this possibility.

If rerolls will be used in the game, EXCON should decide and document the rules governing them before the game begins, including what conditions will result in the award of additional rerolls, and these rules and their intentions should be shared with players during the introduction to the game.

Available rerolls should be marked by some physical item, perhaps a poker chip or a card, that is presented to the team when they have earned a reroll and that is surrendered to EXCON when the reroll is used.

A reroll is what it sounds like: a mulligan, a do-over. Some caveats apply to the use of rerolls:

1. When electing to use a reroll, the new result is final; if it is worse than the initial roll, too bad.
2. When using a reroll, all three dice are rerolled (not just one or a subset).
3. Rerolls can only be used on rolls made by the team using a reroll—that is, a team cannot use its available reroll to force the other team to reroll.

If rerolls are made available, some thought must be given to how many will be made available and over what period or frequency. Rerolls might also be made available as rewards to teams who do something well.

NOTE: The main purpose of the rerolls is to reduce the possibility of an unlucky distribution of rolls souring the experience of a team that is doing a good job in planning but is having bad dice luck. Emphasize ways to award rerolls that try to close the "luck gap" between the two teams.

NOTE: If rerolls are given as rewards, they should reward good planning, presentation, or some other skill demonstrated by the players—not game success determined by dice. The goal of rerolls is to insulate players from some of the randomness of dice, not to reward good luck with an additional tilt toward good fortune.

Here are some possible ways to allocate rerolls (again, whatever method is chosen should be clearly stated to players so that they know how and why they receive rerolls):

- At the end of each day of play, if either team has had two or more successful actions than the other team, the unlucky team gets an additional reroll.
- Each team receives two rerolls for the entire game (all five turns).
- Each team receives one reroll per turn (the authors think this is too frequent, allowing too much rerolling, but we still present the option).
- Any action with an effectiveness target number equal to or less than 8 (that is, any action that has a probability of successful effect of 84% or higher) that fails (and the failure stands and is not rerolled) earns the presenting team an additional reroll that they can use at any time on a later action.
- Any other rewards-based approach to awarding extra rerolls that EXCON can think of—perhaps a reward for successfully contending with a difficult inject, or perhaps a once-per-game on-the-spot award for a particularly outside-the-box but excellent action proposal, or something else. Again, a caveat

is that a reroll should not be awarded for good dice luck; rather, the opposite would be preferable—award a reroll for bad dice luck that ruins otherwise good plans.

2.4.3.8.3.2. OPTIONAL: Single-Roll Adjudication

In the previous edition of this wargame, the default approach to adjudication involved a single roll that determined the effectiveness and outcome of the action, with the execution of the action determined by the narration related to that outcome. That is, it was up to the narrator to describe a failed outcome as the result of either (1) failed execution or (2) a successfully executed action for which there was as a flaw in the players' logic or understanding of the context, such that the successful execution of the action did not lead to intended results. The judges still scored aspects of the difficulty of executing the intended action and its prospects for success if competently executed; this was just all summarized into a single roll. While the current edition includes the two-roll approach as the default, GA8a includes an updated version of the previous edition's single-roll scoresheet, in case the EXCON lead prefers that approach.

Under the single-roll option, EXCON consolidates Sections 2.4.3.8 and 2.4.3.9 to correspond to a single roll, and the roll determines the effectiveness of the action, with execution left to the discretion of the narrator as they describe how things unfolded to deliver the determined success or failure. The outcome roll need only be accompanied by a margin for success for effectiveness (see Section 2.4.3.9.3), and the prepared narratives for different levels of success should also include descriptions of execution.

2.4.3.9. Adjudicating the Effectiveness of the Action

After the outcome of the action's execution is determined, it is time to determine the effectiveness of the action. If the execution was determined to have failed, then the action did not take place and it is automatically ineffective. The action ends without needing a second roll. If, however, the action was at least partially executed, adjudication of the action continues to an effectiveness roll.

2.4.3.9.1. Determining the Final Target Number for the Effectiveness of the Action

The process for determining the effectiveness of an action follows the process for adjudicating the execution of an action. As described in Section 2.4.3.8.1, before the outcome roll for execution was made, the judges should have met to reach a target number for execution and a preliminary target number for effectiveness. (By comparing their individual scores and debate modifiers for effectiveness—parts C and D of the scoresheet—and either reaching consensus or using the middle values.)

There is one more modifier to the target number for the roll for effectiveness: the margin of success from the execution roll. Actions that were fully and completely executed are more likely to be effective than actions that were only partially executed. Table 4 and the engagement scoresheet (part 5) list the success of execution modifiers that should be included in the calculation of a final target number for effectiveness (part E of the scoresheet). Figure 9 provides an example of a completed scoresheet and calculations for the target number for the likely effectiveness of an action. (In the example in Figure 9, the presenter rolled 14 on their effectiveness roll, against a target number of 11.)

2.4.3.9.2. Margin of Success for the Effectiveness Outcome Roll

As described in Section 2.4.3.8.2 for execution, effectiveness outcome rolls will also be governed by a margin of success. Table 5 depicts possible margins of success (or failure) for the effectiveness outcome roll and their interpretation. Remember that outcomes related to effectiveness can also be thought of in terms of MOEs.

FIGURE 9

Example of Completed Judge's Scoresheet for Likely Effectiveness of the Action

Engagement Scoresheet Judge: Joe Turn #: 2 Team: RED Action #: 2

The **Difficulty of Executing** the Action

1 *Score the difficulty of executing the action*

Difficulty of coordinating acting capabilities	Very easy ☐ 0	Easy ☐ 1	Moderate ☑ 2	Difficult ☐ 3	Very difficult ☐ 4
Difficulty of positioning all capabilities	Very easy ☐ 0	Easy ☐ 1	Moderate ☑ 2	Difficult ☐ 3	Very difficult ☐ 4
Difficulty of capabilities remaining for required duration	Very easy ☐ 0	Easy ☐ 1	Moderate ☑ 2	Difficult ☐ 3	Very difficult ☐ 4
Difficulty of capabilities executing activities	Very easy ☐ 0	Easy ☐ 1	Moderate ☑ 2	Difficult ☐ 3	Very difficult ☐ 4
Action's requirements relative to capabilities' capacity	Well within capacity ☐ 0	Within capacity ☐ 1	Max capacity ☑ 2	Over capacity ☐ 3	Well over capacity ☐ 4
Overall difficulty of executing this action	Very easy ☐ 0	Easy ☐ 1	Moderate ☑ 2	Difficult ☐ 3	Very difficult ☐ 4

2 *Sum the scores from 1, then consult the table below and circle the corresponding* **execution score**

Total		Score
0-5	➡	6
6-7	➡	7
8	➡	8
9-10	➡	9
* 11-12	➡	(10)
13-14	➡	11
15	➡	12
16-17	➡	13
18-20	➡	14
21+	➡	15

3 *Determine the* **debate modifier***: After the debate, does the action seem easier or more difficult than originally thought?*

much easier	somewhat easier	the same	more difficult	much more difficult
☐ −2	☐ −1	☐ 0	☑ +1	☐ +2

4 *Determine the* **target number** *for the execution outcome roll by summing the score from 2 and the modifier from 3:*

10 (Execution score) + +1 (Debate modifier) = 11 (Target number)

5 *After the presenter makes their outcome roll, note the margin of success, consult the table below, and circle the corresponding* **success of execution modifier**

Margin of Success		Degree of Execution		Success of Execution Modifier
6 or better	➡	Remarkable success	➡	-2
* 2 to 5	➡	Full success	➡	(-1)
0 to 1	➡	Near success	➡	0
-1 to -2	➡	Partial execution	➡	+3
-3 or worse	➡	Failure	➡	No effectiveness outcome roll is made

GA8

The **Likely Effectiveness** of the Action

A *Score the planning of the action, which could influence effectiveness*

Specification of TA	Ideal ☐ 0	Adequate ☑ 1	Partial ☐ 2	Minimal ☐ 4	Fail ☐ 6
Clarity and logic of intended effects	Very easy ☐ 0	Easy ☐ 1	Moderate ☑ 2	Difficult ☐ 3	Very difficult ☐ 4
Capabilities/activities matched to TA/effects	Very easy ☐ 0	Easy ☐ 1	Moderate ☑ 2	Difficult ☐ 3	Very difficult ☐ 4

B *Score the likelihood of this action achieving its intended effects if it is executed correctly*

How *accessible* is the TA via this action? (awareness)	Fully ☐ 0	Mostly ☑ 1	Somewhat ☐ 2	Partially ☐ 4	Not very ☐ 6
How *receptive* is the TA to this behavior? (motive)	Inclined ☐ 0	Open ☐ 1	Neutral ☐ 2	Resistant ☑ 3	Very resistant ☐ 4
How *able* is the TA to affect behavior? (motive)	Fully ☐ 0	Mostly ☑ 1	Somewhat ☐ 2	Partially ☐ 4	Not very ☐ 6
Magnitude of effect sought	Tiny ☐ 0	Small ☑ 1	Moderate ☐ 2	Significant ☐ 4	Immense ☐ 6
Difficulty of achieving the intended effect with this action	Easy ☐ 0	Somewhat easy ☐ 1	Typically achievable ☑ 2	Difficult ☐ 4	Very difficult ☐ 6
Consistency of this action with effects from maneuver	Fully consistent ☐ 0	Mostly consistent ☑ 1	Unrelated ☐ 2	Slightly contradicts ☐ 4	Under-mined by ☐ 6
Overall likelihood of this action achieving its intended effects if executed correctly	Very high ☐ 0	High ☐ 1	Moderate ☑ 2	Low ☐ 4	Very low ☐ 6

C *Sum the scores from* **A** *and* **B***, then consult the table and circle the corresponding effectiveness score*

Total		Score
0-7	➡	6
8-9	➡	7
10-11	➡	8
12-14	➡	9
* 15-17	➡	(10)
18-21	➡	11
22-24	➡	12
25-27	➡	13
28-29	➡	14
30-31	➡	15
32-35	➡	16
36+	➡	17

Also see **F** *on the reverse side of the page*

D *Determine the* **debate modifier***: After the debate, does the action seem likely to be more or less effective than originally thought?*

much more effective	somewhat more effective	the same	somewhat less effective	much less effective
☐ −2	☐ −1	☑ 0	☐ +1	☐ +2

E *Determine the* **target number** *for the effectiveness outcome roll by summing the score from* **C***, the modifier from* **D***, and the modifier from* **5***:*

10 (Effectiveness score) + 0 (Debate modifier) + -1 (Success of execution modifier) = 9 (Target number)

The narration and scenario team should prepare to narrate possible outcomes for each possible degree of success or failure, as described in 2.4.3.8.2.2.

OPTIONAL: The differences between the actual roll and the target number for different margins of success can be adjusted to reflect that an action is all or nothing, or is high-risk/high-reward, or for actions where partial effectiveness is more likely, at the discretion of EXCON.

OPTIONAL: If optional rules for critical failures and/or critical successes are in place, these will apply to effectiveness outcome rolls, too, and the narrator will need to be prepared for those possible outcomes.

TABLE 5

Margins of Success for Effectiveness and Their Interpretation

Margin of Success	Level of Success in Effectiveness
Beat target number by	
6+	Remarkable success in effectiveness
2–5	Full success/fully effective
0–1	Partial success: approximately 50% of possible effect
Fall short of target number by	
1–2	Near miss: approximately 10% of possible effect
3+	Failure: not effective

2.4.3.9.3. Outcome Roll and Outcome Determination for Effectiveness

With a target number determined for effectiveness, the EXCON lead announces the target number. The EXCON lead should provide more information than just a target number, also noting the judges' overall assessment of the difficulty of executing the action as modified by the degree of success of execution and the probability of beating the target number (see Table 10). Here is a sample statement:

> Based on our assessment of the action and the degree of success of execution, the judges have determined that the target number for execution of this action is [X]. This is because we deemed that this action [should succeed, has a good chance to succeed, could succeed, is unlikely to succeed, or should fail] to deliver the intended effects. This corresponds to a [X] percent chance of success.

The presenting player now makes an outcome roll, by rolling three dice and summing them, as described in Figure 3. This outcome roll total is then compared to the target number.[5]

The results of this outcome roll determine whether the action had the intended effects and to what extent. The EXCON lead announces the margin of success for the effectiveness of the action, and then the narrator offers a brief description of the execution. The narrator should begin by repeating the margin of success and brief narration for the *execution* of the action (repeating the MOP), before repeating the margin of success for the *effectiveness* of the action. The description should clearly restate the **target audience** and the **initial effect**. The narration of effectiveness should focus on preliminary MOE, effects that are initially observable during and right after completion of the action. These may need to be somewhat circumscribed, as other effects on the battlefield (such as the effects from the other team's actions) might modify the overall outcome. Overall outcomes will be fully narrated in Step 5.

> **NOTE:** Narration for the effectiveness of actions in Step 4 needs to strike a balance between giving the players the payoff for their successful actions and the possibility that actions by the other team may affect the ultimate outcome (as to be revealed in Step 5). In general, descriptions focused on initial and preliminary MOE that use phrases like "appears to be working" are safe and provide a hedge against needing to change the overall outcome in Step 5. If the narrator knows that none of the remaining actions affect this same audience or this same portion of the area of operations, then they can be more expansive in their initial narration. When in doubt, err on the side of providing too much information. The narration of the outcome is the capstone of the adjudication process and is where players' hard work pays off (or doesn't). Reward them with detail (hedging only where you know there might be a risk of countervailing or reinforcing effects from later actions).

[5] Section 2.4.3.8.3.1 presents optional rules related to rerolls. If a team has a reroll available and their initial roll falls short of the target number, they should announce their intention to use the reroll and reroll the dice.

Notetakers should record the announced target number, the actual outcome roll, and the key elements of the description of the effect.

Players should also take notes on when and why their action was effective or ineffective. Was it because of bad dice rolls? Could they have improved their chances with better presentations or stronger rebuttals? These notes will help them adjust their plans in future turns of the game.

2.4.3.9.4. OPTIONAL: Presenting Player Offers Self-Critique

To promote learning from failure (and to provide additional fuel for the narrator), if the outcome roll for effectiveness is a failure of some sort, EXCON lead can ask them to explain why they think the action might not have succeeded. This basically invites the presenting player to narrate the reasons their own action might not have worked. This takes advantage of the fact that they have invested considerable time and energy in this action and have (or should have!) thought about things that might detract from effectiveness; it forces them and other participants to think about logical reasons that things might have worked out (other than just a bad dice roll). If the optional rule in Section 2.2.3.1.3 (i.e., having players write down reasons the action might fail to execute or fail to have intended effects) is in use, this self-critique should be even easier. Presenters are not asked for similar input for an action that is at least partially effective: They've already told everyone why they thought the action would be successful.

2.4.3.9.5. OPTIONAL: Hiding Results as a Teaching Moment

As envisioned in this wargame, all players hear the target number announced, and then the presenting player rolls dice and generates an outcome total. Having heard the target number, anyone who can see the dice will immediately know whether the action succeeded or failed, even if they don't know the exact margin of success.

However, there may be circumstances within the scenario that should prevent the results from being immediately apparent for certain actions. This might be because effects are expected to be delayed, or because the effects may not be observable until later. Or the assessment plan associated with the action might be judged to be insufficient to get reliable feedback about the effectiveness of the action, either because it failed to identify good MOEs or because assets necessary to collect or measure those MOEs would not realistically be available.

If realistic circumstances would prevent players from knowing whether or not their action was effective, the EXCON lead has two choices: (1) Tell them anyway or (2) conceal the results. This ruleset prefers a hybrid course of action: combining those two options to create a teaching moment. The first time an action is taken for which available assessment is deemed insufficient to return adequate reporting, when the presenting player makes their outcome roll, the EXCON lead or the narrator (decide beforehand!) should leap forward and cover the dice after the roll but before the roll has been read (or stop the rolling player from rolling right before the roll).

Taking advantage of this dramatic interruption in the expected flow of the game, the EXCON lead or narrator should explain that, in this case, as is often true in the real world, there is insufficient information collected to reveal the results of this action. After using this teaching moment to emphasize the importance of assessment planning and the reality of fog of war in the information environment, the EXCON lead or narrator can then explain that, even though in real life we would not yet know the results of this action, for the purposes of the game we are going to reveal it. The dice could then be revealed and counted, and adjudication could proceed as described above.

This teaching moment can then be referenced as a reminder every time there is an action that might not be fully observed or assessed, as encouragement to include strong assessment plans, and perhaps as a further reminder for players to plan their future actions based only on things they should know, rather than including information that is revealed in Step 4 to allow adjudication.

OPTIONAL: If there is an action taken on one turn (and which would take place during the time bounds of that turn) that should not have effects until a subsequent turn, EXCON may elect to delay the outcome roll until that later turn. In that case, the action would still be presented in the turn in which it takes place, and the judges would still determine a target number, but the actual rolling of the dice and determination of the outcome would be held until Step 4 of the turn in which the effect should take place. In that turn, just the outcome roll portion could be inserted into the action sequence in a position deemed appropriate by EXCON, and results would be reflected in Step 5 of that turn.

2.4.3.10. Repeated, Recurring, or Ongoing Actions

Circumstances within the game may lead to a team repeating an action. This might be a new instance of the same action, an ongoing effort that was previously successful, or an ongoing effort that was previously unsuccessful.

The acting team will present their action regardless. However, EXCON must determine how they wish to adjudicate it. If this is a new instance of the same action, then it should be treated like a new action in all respects, because it is a new attempt to execute the action. However, if the action is ongoing, then alternative adjudication may be appropriate.

If an action is ongoing but has not yet been successful, then it should be subjected to the full adjudication process (presentation, matrix discussion, scoresheets, target numbers, outcome rolls, etc.). An ongoing but unsuccessful action may be subject to changes in target numbers based on previous outcomes, either a bonus for cumulative effort/effect or a penalty for repeated ineffective intervention with the same audience.

If an action is ongoing and has been successful, then EXCON must decide whether to simply allow it to continue successfully or to require a new adjudication and rolls. This determination can be made at the conclusion of the substep in which the non-presenting team rebuts (Section 2.4.3.4). If not much has changed since the acting team's action began and the opposed team has no new or particularly compelling reasons why the ongoing action should not continue to succeed, then EXCON may simply allow that it continues at the same level of success. This would prevent the presenting team from facing "double jeopardy" and being exposed to multiple dice rolls (and chances of failure) for something that has already been determined to be successful.

If, however, circumstances have changed that would in any way call the continued success of the action into doubt, then the action should be re-adjudicated and new outcome rolls should be made. The judges may choose to give a roll bonus as appropriate for continuing a successful ongoing action even if circumstances have changed such that it is no longer a slam dunk.

Optionally, an ongoing action can just be subject to an execution outcome roll to determine whether the side's information capabilities continue to execute the action successfully, with the effectiveness of the action remaining the same if the degree of performance (MOP) remains the same.

NOTE: Requiring teams to present their ongoing actions for approval and for adjudication helps maintain a check on resources. The EXCON members role-playing S3s may quickly re-approve continuing actions, but will consider that ongoing action and its resource requirements against the overall resource draw of the team for the turn, and check against capabilities that have been exhausted or attrited (see Section 2.5.4.4).

2.4.3.11. Repeat Activities as Necessary

The substeps described from Sections 2.4.3.3 through 2.4.3.10 should be repeated for each action until all approved actions for the turn have been presented and resolved.

2.4.3.12. Record All Results and Prepare for Step 5

EXCON members should have been taking notes on everything throughout, but as the process moves into Step 5, it is particularly important to make doubly sure that outcomes from all actions have been recorded, as have the final positions and dispositions of elements on the game board.

2.4.4. Outputs from Step 4

Step 4 should produce notes on the actions completed, the degree of success or failure of their execution and effectiveness, and their outcomes as described by the narrator. Final dispositions of actions as adjudicated on the game board will be presented and determined as part of the summary narrative in Step 5. Output for filing and recording should include all the completed scoresheets for all of the actions.

2.4.5. Time Allowed for Step 4

Step 4 has the greatest number of activities and outputs of any step within the wargame, and, even though the procedure is fairly structured, it has the potential to take longer than anticipated. Experience shows that the average action takes fully 30 minutes to adjudicate, with six actions taking approximately 3 hours of play time. Adding in appropriate breaks and considering the extra time needed to complete Step 4, taken together, Steps 4 and 5 should be allocated an entire half-day of play time, either an entire morning or entire afternoon session.

2.5. Detailed Conduct of Step 5: Results and Reset

2.5.1. Overview of Step 5

Step 5 should begin immediately on conclusion of the resolution of the final action taken during Step 4 and should occur while both teams are still in the engagement room. The players get to relax for a few minutes while EXCON completes Step 5, recording all outcomes, narrating the final effects and outcomes of all actions, and determining their impacts on the overall progress of the scenario. Step 5 concludes with an in-stride after-action report (AAR) taking advantage of any teachable moments and making observations or suggestions that could improve the game in the next turn.

2.5.2. Inputs to Step 5

Most inputs should already be present: projection capabilities, game board, results of all outcome rolls from all actions, notes for narrating the final outcomes of all actions, and tokens or markers to note the final dispositions of actions on the game board. Additionally, the anticipated results for maneuver for the turn should be ready and ready to present. The EXCON lead should have prepared notes for the in-stride AAR.

If this is the final turn, inputs include the preparation of any awards that will be presented.

2.5.3. Player Activities During Step 5

Players should be attentive and take notes. Players also should get fired up: either to follow up their successes and stick it to the other side, or to lick their wounds and get back on their feet. The next turn is a new day! In the final turn, players should stand by with bated breath waiting for the final adjudication of the wargame and the presentation of awards.

If players have an urgent question or if something in the storyline is unclear, they may bring it to the attention of EXCON during Step 5. Such questions should either be addressed to the EXCON lead or the narrator as appropriate. This should happen at the end of Step 5, during the in-stride AAR.

2.5.4. EXCON Activities During Step 5

EXCON members, specifically the narration and scenario team, should quickly meet and finalize the outcomes and effects of all actions, paying special attention to where later actions may have interfered with the expected results of earlier actions (for example, where an early action with a successful effectiveness roll might be diminished by a later action by the other team with a successful effectiveness roll seeking a contradictory effect). The narration and scenario team should also determine whether any of the actions for the turn have any consequences for the projected trajectory of maneuver outcomes and the storyline for the turn and, if so, quickly identify these changes. Such changes should not be a surprise, and adjusting for them should be quick and easy. Anticipating such changes should be the responsibility of the narration and scenario team.

The narrator then presents a summary narrative of the turn, covering all of the following, in sequence:

1. Review of initial conditions for the turn
2. Review of each action from the turn, noting the margin of success of its execution, the margin of success of its effectiveness, and the actual effects and consequences from the action
3. The impact of actions on the overall maneuver or operational progress for the turn.

The review of the initial conditions should be very similar to the summary that began the turn in Section 2.4.3.1, but the narrative for each action and the overall summary of impact of the actions on operational progress merit additional discussion.

2.5.4.1. Narrating the Final Outcomes of Each Action

In Step 4, the narrator described the margin of success of the execution of the action, announced the margin of success of the effectiveness of the action, and offered some preliminary observations about the possible effects of the action. This review of each action should repeat that information and add more detail, extending the description of effects to include the overall impact of the action during the time covered by this turn.

NOTE: Finding the right balance between narration of effects in Step 4 versus what is finalized in Step 5 can be tricky. Ideally, the narrator should provide as much detail as possible during Step 4, right after the outcome roll for each action, without risking describing results that will then be at risk of being changed by later actions. Err toward providing more information in Step 4 unless you know that later actions might affect the same actors or audiences.

Remember that we call the position *the narrator* and describe the activity as *narrating* because it involves storytelling! While the relationship between the outcome rolls and the target numbers is immediately apparent, and participants have already been given a margin of success or failure for the action, it is still up to the narrator to make clear what the numbers mean and make a declarative and/or narrative determination of the outcome and its consequences. These should be based on the interpretation of the margin of success given in Tables 4 and 5 (and in GA7 and GA8), the nature of the action, and the descriptions of success and failure offered in the action presentation (the narration and scenario team should not be married to those descriptions, but they are a place to start). Possible descriptions of success or failure for each action are something that EXCON and the narrator could or should have given some consideration to during slack time in other steps so that they are prepared to describe outcomes (see Section 2.4.3.8.2.2).

OPTIONAL: If used, astounding successes and critical failures (see Section 2.4.3.8.2.1) should be described in hyperbolic terms and should (provided the presented description was at all reasonable) meet or exceed the "what success looks like" or "what failure looks like" outline as described in the presentation of the action.

This verbal description of the final outcome of actions should be accompanied as appropriate by additions of icons or symbols or other adjustments to the game board. Tokens, markers, or flags could be added to denote effects; effects could be written in grease pencil next to tokens for affected audiences; tokens or figures representing audiences could be laid down or rotated from vertical to horizontal in their stands to denote an effect—however it is done, effects should be depicted on the game boards.

The EXCON member acting as narrator should remember to role-play as the narrator for this story and make the descriptions of the outcome dramatic, definitive, and entertaining. Remember that the dice do not speak for themselves, and the narration of the outcomes is the culminating moment of both teams' efforts for the turn. A big part of the excitement of the game hinges on the throws of the dice and the description of what happens by the narrator. This is how the players know that their team's actions mattered! Keep in mind that the narration needs to remain within the bounds of reality. Where feasible, narrators can refer to the presentation's "what success looks like" or "what failure looks like" outline as a guidepost. But narrators should feel free to correct these descriptions unilaterally if the presenters were over-the-top or too modest, as well as come up with other descriptions of the effects of the action (provided that they stay within the bounds of the overall storyline).

A narrative explanation of an action outcome should cover several things:

1. Restate the action with a clear statement of the TA.
2. Repeat the target numbers, the dice rolls, the margin of success achieved for execution, and the margin of success achieved for effectiveness.
3. Deliver an account of MOPs, based on the execution outcome roll and margin of success. The description of MOPs should include what the presenting force was actually able to do, even if

it fell short of what was planned. If any capabilities were exhausted or attrited in relation to this action, repeat that, too.

4. Describe the outcome in terms of **the effect or impact on the TA and their change in attitude or behavior**.
5. Finish with a connection to the overall operation and any impact (or lack of impact) on the operation.

For each possible level of success or failure, the narrator should have a plan for what to report depending on dice roll result. A narrator worksheet is included in the game materials.

For **full effectiveness**, the narrator should use the presenter's explanation of "what success would look like" as a starting point. Preparing narration for full effectiveness is relatively easy, as the presenters will have described what they want to happen. Narration need not be an exact read-back of what they intended to happen, however. If what the presenter requested is somewhat unrealistic, make notes for and describe reasonably successful effects based on the action. Also, consider additional effects (the impact on overall maneuver for both sides, for example, or secondary impact on GREEN) and be prepared to describe them.

OPTIONAL: **Critical failures** and **astounding successes** (see Section 2.4.3.8.2.1) are fairly easy to describe. For an astounding success in effectiveness, give the presenters everything they asked for and more—imagine and describe the most fortuitous outcome from that action! And remember to protect the storyline: The action can be an astounding success with immense effects, but try to keep those effects from pushing the storyline out of bounds. For a critical failure, visit the full wrath of Murphy's Law on the presenting team: Perhaps a blowback of effects that causes exactly the opposite of what was desired. This may be attributable to an accident—for example, a population that was encouraged to move south actually moves north because of a translation error (if moving north would interfere with the acting side's concept of maneuver)—or to a deception exposed and exploited, provided that the explanation doesn't overly damage the storyline.

TIP: Don't let an outcome totally break the storyline! Narrators should remember that there is a default storyline that is pre-scripted for the scenario, and the scenario has intended left and right bounds of extreme BLUE or extreme RED success that more or less keep the storyline under control. Narrate outcomes, even extreme success or failure, to keep the storyline within those left and right bounds. One of the big variables you can allow to change based on action outcomes is timing.

Successful actions can cause delays in one side's progress, and such delays are much easier to accommodate in a dynamic storyline: If something is delayed, it still happens, just later. If one side really delays the other, then the intended storyline can still play out, just more slowly. The delayed side may ultimately be deemed to have failed if they are sufficiently delayed, and that can happen without having to change the core storyline at all.

After outcomes for each action have been narrated, they should be added to a running list of all actions that have been attempted during the game that includes the degree of execution success of each action, the effectiveness of each action, and summary notes about the impact and outcome of each action. Not only should the narration and scenario team maintain such a list, but **the list of all actions attempted and their outcomes should be publicly displayed in the engagement room**. This might be as a slide on a monitor or projected display, or it might be on a whiteboard, on a chalkboard, or on a butcher paper pad or easel. A caution: If this list of actions and outcomes is handwritten, care should be taken to write clearly and legibly.

2.5.4.2. Presenting the Summary of Impact of the Actions on the Course of Operations

After narrating the effects of each action, the narrator should summarize the overall impact of those actions on maneuver progress and other operational progress for the turn. **The core goal of this summary is to take the results of the information actions from the teams for the turn and put them back into the context of the broader operation.**

The narrator should take pains to make it sound like successful and potent actions are having an effect on the context and on maneuver, even if maneuver plans remain within tolerance and are pretty much on-script. In the wargame, as in real life, perception is reality. If players believe that their actions have affected the shape of the operation, then they will feel much more empowered and consequential. The narrator should adopt a role-play attitude and should describe both the major information and maneuver activities and other operational progress dramatically.

The presentation in Step 5 is a summary of everything that has happened (information and maneuver) during the turn. This is different from the Step 1 presentation, which is a summary of the current state at that time and projection of what is scheduled to happen during the turn to come.

After concluding the summary of actions completed and providing the update to the storyline covering what has happened during the turn, EXCON should caution players about which of the results and consequences from the turn's actions are "scenario knowledge"—things the information OPT would be aware of in real life—and which are "player knowledge"—things that the players got to see adjudicated with the God's-eye view of the engagement in front of EXCON, but that in the context of the scenario remain shrouded by the fog of war and the limitations of assessment. The phrase "what happens in the engagement room stays in the engagement room" might be a good reminder of this distinction.

2.5.4.3. In-Stride After-Action Report

Following the conclusion of the narrator's summary of the impact of the turn's actions on the overall flow of the scenario operation, EXCON lead leads an in-stride AAR. This AAR has two core purposes: first, to capture useful learning—insights from teachable moments, observations about good tactics, techniques, and procedures (TTPs), instruction on how certain processes or capabilities actually work, etc.—and, second, to make suggestions or corrections that will improve the quality of game play or the game experience in the next turn. This might include suggestions about how to frame counterarguments, or a request to dial back rivalry and aggressiveness and dial up mutual courtesy (see Section 2.6.3). Within those general guidelines and as time is available, EXCON lead can run the AAR to suit their preferences, possibly taking questions from participants or inviting observations from other EXCON members, observers, or players.

During Step 5, members of EXCON should also be preparing for Step 1 of the next turn. Materials for Step 1 of the next turn should already be pre-prepared and follow a pre-prepared storyline but may require adjustment based on the rolls and outcomes of the current turn. This may need to happen simultaneously with the summary of the turn's results by the narrator, and so it may need to involve other personnel (perhaps in collaboration with those who will be presenting the Step 1 update).

Step 5 requires the narrator to give the summary narrative/storyline for the turn. It also requires EXCON personnel from the narration and scenario team to quickly consult with each other and adjust the storyline for maneuver based on information action results.

Finally, the narration and scenario team should be listening to the narrative provided by the narrator and preparing any adjustments necessary for the update presentation in Step 1 of the next turn that this necessitates. Five to nine EXCON personnel are required for this step, depending on how many of the roles are included and which, if any, are either dual-hat responsibilities or are assigned more than one EXCON representative.

If this is the final turn of the game, the narrator will not only summarize the turn but will also narrate the conclusion of the conflict, indicating the winning side and the winning team (which need not be from the winning side). EXCON will present any awards at this time. (See Section 2.5.5.1.)

2.5.4.4. Determination of Capabilities Exhausted or Attrited During the Turn

Once the players have been released from the turn's narration and the in-stride AAR, EXCON makes final preparations for the next game turn (unless this was the final game turn). In addition to plotting where the storyline is relative to different possibilities (see Section 3.4.5) so the correct initial positioning brief can be given to the teams for the next turn (see Section 2.1.4), EXCON needs to determine whether any capabilities have been exhausted or attrited during the turn and thus will be unavailable in the following turn (see Section 2.4.3.8.2.2). Capabilities might be unavailable because they were narrated as damaged or destroyed based on a critical failure or extreme outcome on the execution outcome roll (see

Section 2.5.4.4), because a capability has been repeatedly requested and denied and the S3 or narrators has determined that capability will never be available (see Section 2.3.4.2), or simply because a capability is being overused and has become a crutch for the team, and EXCON wants to push them to try something else (perhaps noted by the S3 during Step 3, or perhaps based on EXCON observations up until this point). Note that capability exhaustion and attrition might be temporary conditions: Exhausted capabilities might rest and replenish, and attrited capabilities might reconstitute or be replaced; this substep allows for EXCON to reconsider the availability of capabilities removed from availability in previous turns. In fact, such a determination might be made at the outset. For example, a team that has heavily relied on and overtaxed their military information support operations (MISO) capability might be told that the capability is exhausted and is unavailable for one turn; or a corrective inject (see Section 2.6.1.2) might note that all electronic warfare capabilities are needed for a priority mission at a higher echelon and are unavailable for the next turn.

NOTE: Any determinations regarding capabilities that are not available to a team for the next turn should be added to the Step 1 snapshot scenario and situation update and should be emphasized in that briefing so that players are aware of and understand the change.

2.5.5. Outputs from Step 5

Step 5 produces an extended narrative of the conduct and consequences of each action from the turn, as well as an updated situation to inform the next turn, including possibly durable changes to the game board (such as changes to available capabilities) and adjustment of the expected track for maneuver. Step 5 also produces notes and records for the AAR and subsequent turns. As noted at the end of Section 2.5.4.1, output should also include a cumulative list of all actions attempted in the game and their outcomes, including execution, effectiveness, and a summary of overall result. This list should be publicly displayed in the engagement room, either as something displayed on a monitor or by projector or as a handwritten list on a whiteboard, chalkboard, or butcher paper.

In Step 5 of the final turn, outputs include various awards (ideally accompanied by certificates, but certainly by good feelings and good cheer; see Section 2.5.5.1).

2.5.5.1. Outputs in the Final Turn of the Game

The game will end on a Step 5 of the final turn. At that time, the narrative summary of the turn should be expanded to cover not only the turn but also the trajectory and course of the whole operation, as well as the outcome. A winning team should be identified based on the extent to which they were able to support their side's concept of maneuver and impact the overall trajectory of operations.

NOTE: The winning team need not be from the winning side. One side in the conflict may be much stronger and destined to prevail in the conflict, but excellent performance by the weaker side's team might make that victory take longer or be less complete, and that should be acknowledged and rewarded. For example, even if the BLUE operation succeeds, the RED team might have made it so much more difficult on the BLUE force that they "won" in the information environment.

Ideally, all members of the winning team should receive a winning certificate. Other awards might also be presented, including humorous ones. Awards can be to individuals or for actions. Awards for actions will go to the presenting player for that action, but with recognition that others from the team will have contributed to the development of that action.

Examples of possible awards include best presentation, best rebuttal, best presentation of a failed action, action with biggest impact on maneuver, action most emblematic of the principles of the information warfighting function, EXCON's favorite action, most embarrassing proposed action, action furthest outside the box, least likely to be friends with the Judge Advocate General, and "better lucky than good" (for an action with high target numbers that succeeded anyway).

OPTIONAL: The final narration, declaration of the winner, and the presentation of the awards may be held until the wargame AAR.

2.5.6. Time Allowed for Step 5

Step 5 should usually not take very long to execute, as results should fit within one of the predetermined tracks (storylines) for overall operational progress, and many of the effects and results will have been previewed during the adjudication of actions throughout Step 4 and anticipated as possibilities by EXCON based on their awareness of what was being presented in Step 3. However, if outcome rolls have been extreme or if EXCON has been surprised by a team's actions, more time may be required.

Initial recommended allocation of time for this step is 30 minutes, with the understanding that it may be used as buffer for an overflow from Step 4.

OPTIONAL: If Step 5 takes place at the end of an exercise day, EXCON may immediately segue into Step 1 of the turn for the following day, to allow players to work on their Step 2 preparations overnight.

2.6. Troubleshooting Problems During the Game

Realistically, problems of some sort will arise during the game. Players may take an action that surprises EXCON or an S3 role-player, or an extreme result on an outcome roll may dictate catastrophic failure or astounding success with impacts extending beyond prepared storyline options. When such a situation occurs, keep calm and carry on.

Two principles from other contexts offer guidance. First, "the show must go on." This wargame is not entirely unlike a live theater event, and so the game must continue without major interruption, even if that requires generous improvisation from EXCON. EXCON personnel should put their heads together, quickly talk it through, then decide on how to proceed. Second, "when in doubt, roll and shout." If something cannot be reasonably and fairly adjudicated (it is outside the scope of the scoresheets and parameters established by this ruleset, or it should be treated as genuinely contingent, or an arbitrary ruling would unfairly favor one team over the other), then the EXCON lead (or the narrator) should declare a target number, make an outcome roll, dramatically announce the meaning of the roll, and proceed with that outcome as the final adjudication of the matter.

2.6.1. Dealing with Turn Outcomes Getting Out of Bounds

As envisioned (and described in Section 3.4.5), the wargame has a range of possible outcomes for each turn, driven primarily by predetermined maneuver outcomes but capable of being affected by information activities along three tracks or storylines: (1) an expected direction of progress, (2) progress based on RED information activities outperforming, and (3) progress based on BLUE information activities outperforming. If teams and dice are relatively balanced, turn-by-turn progress of the overall operation should remain within the bounds envisioned by the scenario. However, they might not. If progress threatens to get out of bounds, there are two tools to address it, described in the next two sections: skipping or repeating a turn and using corrective injects.

2.6.1.1. Skipping or Repeating a Turn

One way to get the game back in bounds might be to skip a turn or repeat a turn in the planned sequence. Consider a scenario in which one force is the aggressor and is looking to skip turns/phases and get ahead, while one force is the defender and is looking to slow (or completely derail) the progress of the other side. If the aggressor force makes so much progress in one turn that the planned bounds for the next turn will not accommodate the results of the aggressor's actions, but the planned bounds for a future turn will, consider skipping to that turn. Or, if the amount of progress implied by even the lower bound is not plausible given the disruptions caused by information activities, repeat the turn, asserting that the attacking force basically made no progress and needs to keep trying before they can progress to the next phase. By *repeat the turn*, we do not mean a do-over—the turn still happened, and the outcomes of all actions influence the scenario, but not in a way that is significant enough to advance the progress of the scenario. Players are still *told* that it is the next turn (the number of the repeated turn +1), but EXCON has determined that the next turn's Step 1 briefing on the state of the operation will be functionally the same as for the prior turn because effectively no progress has been made. The next turn is essentially a repeat of the previous one.

Using either turn skipping or turn repeating might force a change in the total turn count. This would either reduce the game length (aggressor wins early) or require an extra turn (defender significantly delays the inevitable or perhaps even prevails). The EXCON lead will ultimately make the call. If the decision is to change the number of turns, this should be clearly explained to the players.

NOTE: The idea of skipping or repeating a turn could be interpreted in at least two ways, and both could be valid in this wargame. The first underlays the intent of the preceding paragraph and just means that the operation does not progress along the storyline as intended because the actions of the (probably defending) force have prevented the operation from moving forward to or toward the next phase as scripted and the teams must fight another turn under pretty much the same scenario circumstances. In that case it would still be a new turn, but the Step 1 update of conditions would be virtually identical to the Step 1 update from the previous turn.

The second possible interpretation is a "do-over" or "mulligan" where teams are instructed to forget what happened in the turn and start over and try again. This latter interpretation should be avoided if possible, but might be necessary if one or both teams fail to effectively execute most of the steps of a turn or fail to offer actions in Step 3 that are sufficiently well developed to be approved, and everyone really just needs to try again.

2.6.1.2. Intentionally Causing Problems During the Game: Scenario Injects

It wouldn't be a wargame without the possibility of surprises, which in wargames are termed *injects*. We envision the possibility of three kinds of injects: *variation injects*, *challenge injects*, and *corrective injects*.

Variation injects (or "white noise" injects) simply stimulate the teams to be dynamic and responsive. By presenting an unforeseen event (or just chatter about a routine event or even a non-event), the team might have to adjust its plans and actions to take account of the change in context. Use of variation injects should only be necessary if the opposing teams are not providing much unanticipated stimulation, either because their actions aren't surprising or interesting, or because their actions aren't succeeding (bad dice luck).

Challenge injects make the game harder for one or both sides. If things are going too smoothly, challenge injects don't just provide something unexpected to stimulate the players—they provide something *difficult* to stimulate the players. This could be something like a significant civilian casualty incident. Challenge events are generally one-sided, in that while they might require a change in plans by both sides, they are only bad/challenging for one side. To balance this, the total number of challenge injects planned for the game should be even for both sides, even if they don't occur on the same turns (for example, RED could face challenge injects on turns 2 and 4, and BLUE could face challenge injects on turns 1 and 4). Challenge injects should be balanced, unless they are being used as corrective injects.

Corrective injects are only used if the need arises. A corrective inject adjusts for one team totally outperforming the other team, or the luck of the dice brutally disadvantaging one team but not the other. The intent of the wargame is for it to be fair and for success or failure to be real possibilities depending on play, but too astounding of a degree of success by one team too early in the game could push the pre-scripted material out of bounds and make things unduly difficult for the narration and scenario team. One way to get the game back in bounds might be with a corrective inject, which is basically a challenge inject that is not paired with a balancing challenge inject for the other side.

Injects of any kind should be announced during Step 1 of the turn on which they occur to allow the team to properly account for them in their planning. As appropriate, both teams should receive information about the inject based on what that side would know.

Injects should be prepared on inject cards, which can then be given to one or both teams as they prepare their turn (note that some inject cards will require a RED and a BLUE version of the card, as events may be perceived differently be the different sides). Ideally, the scenario should include a whole deck of inject cards of each type in order to allow EXCON some preplanned flexibility. This ruleset does *not* advocate random draws from the inject deck; the deck format is just to provide variety in what is available, not to add an unnecessary (and potentially risky, from the EXCON perspective) random element.

See Section 3.4.6 for more on preparing scenario injects.

2.6.2. Dealing with Uncertainty About How Things Actually Work

The U.S. armed services have a wide range of platforms and capabilities; if partners and adversaries are included, the world's military capabilities are many and varied indeed. It is unrealistic to expect EXCON to know everything about every capability available in the operation and performance specifications of all of them. While someone on EXCON should have some knowledge of the core capabilities assigned to the two sides (see Section 3.4), that knowledge cannot be exhaustive. At some point in a game, a question about the operation or effect of a piece of gear, a platform, or a capability will come up that no one on EXCON is immediately prepared to answer. Such uncertainty might pertain to a process or a regulation, too.

How to respond to urgent uncertainty depends on when it comes up. If a question about a capability comes up during Step 1, 2, or 3, saying, "I don't know but I'll get back to you" is perfectly reasonable, provided that someone from EXCON can then attempt to run the needed facts to ground. However, if the uncertainty comes up during Step 4 (engagement), a response may be needed in order to adjudicate the current action, and a delay for research may not be possible. Moreover, such uncertainties are reasonably likely during engagement, when one side has been making an assumption about the operational parameters of a capability that the other side (rightly) calls into question during their rebuttal.

In such a situation, where there is uncertainty and urgency, the EXCON lead should make an adjudication, either making an expert judgment or declaring it a 50/50 toss-up and making an outcome roll, in which a roll of 11 or higher favors one side's interpretation and a roll of 10 or less favors the other. Regardless of how the uncertainty is resolved, every effort should be made to find a source (or a SME) that can provide the correct answer, and that answer should be shared with all participants for education purposes during that turn's in-stride AAR. Note that whatever was decided at the time is how the capability worked in the scenario universe during the relevant turn and action. Even if the wrong adjudication was made, the story will not be rewritten to accommodate what the adjudication should have been. The past is the past, and the game only moves forward.

2.6.2.1. Dealing with Something That Will Not Work the Way Someone Intended

A related challenge is when one team makes an assumption about how a capability works but is unambiguously mistaken, or presents an action that requires a capability to travel or reach an implausible distance. If something is genuinely impossible, it would be best for the S3 (with support from other members of the narration and scenario team, such as the reality master) to reject the action and prevent it from being presented. If a physically impossible action is presented in Step 4, it will be up to EXCON lead to determine what happens. Possible solutions include the following:

1. on-the-fly adjustment by the presenter, curtailing the scope of their action so that it is consistent with reality (and judges adjust their scoring appropriately)
2. the action is deemed so impossible that it automatically fails
3. even though it strains the laws of physics, the action is allowed to proceed as an artificiality of the game, but judges increase the target number for execution of the action (and perhaps by even more than a few extreme scores within the scoresheet would indicate); in this situation, EXCON lead should clearly announce that some of the elevated target number is due to the implausibility of the action.

2.6.3. Dealing with Players Who Game the Game or Stress the Bounds of Civility

Despite guidance to the contrary in the *Player's Guide* and in the introductory presentations to the players, some players' competitive drives can lead them to behavior that runs counter to what is ideal for the success of the wargame as a training event. For example, in this and other wargame contexts we have seen participants "fight the scenario"—that is, engage in extended disagreements about what is fair or realistic—or "game the game"—that is, propose actions or moves that would not make sense for someone in their assigned role in the game scenario but might give them or their side an advantage within the structure of the game. Another form of challenge includes a player becoming hostile and argumentative toward members of the opposing team and stretching the bounds of civility, mostly likely in this wargame during Step 4, which by its nature creates an adversarial context.

One of the keys to avoiding challenges like this in the first place is in the statement of expectations at the outset of the game, and in the social climate established throughout the exercise and the wargame. EXCON and other leaders can display attitudes and lead through examples of the "right way" to play and to engage with the game and other participants. Should that prove insufficient, the Marine Corps has a strong and healthy leadership tradition of taking errant subordinates aside and counseling them regarding their behavior.

3. Preparing for the IWX Wargame

This section contains notes, observations, and suggestions related to preparing the wargame prior to execution. Many of these issues and activities should be considered as part of the planning conferences and other preparation efforts for the IWX of which the wargame will be a part.

3.1. Selection and Assignment of the Players

As briefly noted in Section 1.5, this game is designed for teams of 6–12 players. Playtesting has revealed that as few as 3 engaged and experienced players (experienced with information capabilities and planning, not necessarily with wargames or with this wargame) can form a sufficient team. Teams larger than 12 should be avoided, if possible, as they start to become unwieldy and hard to manage (both from a team leader perspective and from an EXCON/training perspective). They also risk wasting time with excess players being unengaged, which can diminish the experience. See Section 5 for a discussion of ways to scale the IWX Wargame to accommodate more players.

Available IWX participants can be divided into players on the two teams by any expedient manner. Players/teams can be assigned by exercise staff, or exercise staff can choose two team leaders and allow them to choose teams (perhaps using some variation of "schoolyard pick 'em," or something more akin to an NFL draft).

Allowing team leaders to pick teams is certainly a fair approach to begin a competitive game, but it may not be optimal for targeted training and education purposes. The following are some considerations that might cause EXCON selection of teams to be preferred and criteria that might be used:

- There might be certain personnel that will deploy soon in a certain role, and having them practice that role during IWX would be to their benefit (for example, making sure they are on the BLUE team and have a certain role within that team, such as team leader).
- Ideally, each team will have a distribution of capability backgrounds and expertise; if there are two public affairs specialists and three psychological operations specialists among the participants, each team should have at least one of each. That might happen with participant-picked teams, but it can be assured with EXCON-assigned teams.
- Ideally, levels of seniority and experience will be divided between teams. A situation where all the senior personnel with practical experience are on one team and the other team is composed of exclusively more junior personnel might both make the wargame less fair and limit opportunities for within-team mentorship.

3.2. Requirements for EXCON Personnel

The wargame requires at least 4 EXCON personnel to execute. Optional additional support could engage 9 or more additional personnel—see Table 6 in Section 3.3. These individuals should meet these general requirements:

- Be available to participate in the full span of IWX and available for EXCON rehearsals/training (as detailed in Section 3.2.4).
- Have some experience with planning information activities and with various information capabilities. IWX and information working groups or OPTs, simulated or otherwise, tend to have acronyms flying thick and fast, and EXCON personnel must be comfortable with the vocabulary and the flow of such efforts.
- Have some experience or at least familiarity with information capability training events or exercises like IWX and some familiarity with wargames (either as a hobby gamer or though observing or participating in military wargames).
- Be patient, detail-oriented, and willing to read and reread rules.
- Be able to work collaboratively with others.
- Be able to stay within their assigned roles and not default to teaching and instruction (unless their role during the game explicitly involves teaching and instruction) or losing objectivity

(becoming part of the team is not fair or helpful to achieving the learning objectives).

In addition, some of the specialty roles require additional specific skills, as described in Sections 3.2.1–3.2.4.

3.2.1. Requirements for the Operations Officer (S3) for Each Side

The role-players for the operations officers/S3s (or RED equivalent) are absolutely critical. These must be veteran and expert personnel who are capable of challenging the players; forcing them to defend, explain, and justify their proposed actions; and giving them a realistic "OPT lead briefing the S3" experience. In addition to the required experience and force of personality, these role-players must also be very well-versed in the details of the scenario and the operation. The S3 role-players should also be part of the narration and scenario team and should be drawn from the cadre of exercise planners who worked to prepare the scenario for the event.

The S3 role-players are responsible not only for giving the players an appropriately hard time, but also for ensuring that the correct number of well-planned and well-presented actions arrive in Step 4.

It is imperative that the same role-players represent these critical roles throughout the entire game, not just for continuity, but for consistency in the filtered view of the scenario and operation that is provided.

It is also imperative that the S3 role-players protect their role as hard-nosed S3s. They should avoid too much hallway jocularity and should ideally not be dual-hatted as a senior mentor. Their role is the boss and the evaluator, not the coach.

Experience with the game has revealed this as one of the most challenging but also most important EXCON roles. Poor S3 execution can result in substandard actions reaching step 4; distortions of the scenario reality, when role-play and improvisation have gotten carried away; and disconnects between the players' understanding of the scenario and the baseline maintained by the narration and scenario team. For more on the kinds of pitfalls that await an unwary or ill-prepared S3 and the problems that can result, see Section 2.3.4.4.

The S3 role-player should not be a distinguished visitor, but should instead be drawn from the narration and scenario team: Familiarity with the scenario and intended storyline is the single most important qualification for an S3 role-player.

A list of S3 role-player duties by game step is provided in *GA5: S3 Role-Player's Responsibilities.*

3.2.2. Requirements for the Judges

Judges may be either civilian or military and should be relatively senior. Judges should be well-versed in the Marine Corps Planning Process or the Joint Operation Planning Process.[6] Judges must have substantial experience, bordering on expertise, with the information warfighting function and one or more of the information capabilities. Judges must have discriminating judgment and the ability to discern what is practical from what is impractical and what is easy to accomplish in the information environment from what is hard.

Judges should be at least passingly familiar with the scenario, since team actions must be evaluated in context. A panel of judges who arrive at scores by consensus is preferable to a single judge, and a panel of three judges is recommended. Judges must be familiar with the game's scoresheet (or willing to gain familiarity through EXCON training/rehearsals, as described in Section 3.2.4).

A list of judge responsibilities by game step is provided in *GA3: Judge's Responsibilities.*

3.2.3. Requirements for the Narrator

The narrator should have a good baseline of expertise related to the information environment, a high level of mastery of the game rules, and close familiarity with the scenario (they are the heart of the narration and scenario team, and ideally would have been part of the cadre that developed the scenario for the exercise). However, these things alone are not sufficient to fulfill this role. The narrator needs to have a certain level of something like the "gift of gab" and a certain degree of quick-thinking creativity. Individuals with the right stuff for narration might

[6] MCWP 5-10, *Marine Corps Planning Process*, U.S. Marine Corps, as amended August 10, 2020; and Joint Publication 3-0, *Joint Operations*, Joint Chiefs of Staff, June 18, 2022.

well have a background that includes improvisational theater, might have experience as a "dungeon master" or "game master" for tabletop role-playing games, or could just be an accomplished and captivating storyteller. These are not the only ways to qualify, but note that the needed qualities should not be assumed to be universal among otherwise intelligent and accomplished professionals. There is an art to this, and not everyone has the necessary talent. For example, look for someone who artfully tells stories that are on topic and germane to the discussion at hand rather than someone who frequently offers long personal tangents and digressions. It is essential that the right person be found for this role, and that it not be considered a "prestige" position offered to the most prominent member of the narration and scenario team, for example. New narrators should be pushed and tested during rehearsals to make sure they have the right stuff.

A worksheet to help the narrator during Step 4 is provided in *GA7: Narrator's Preparation Worksheet.*

3.2.4. Training and Rehearsals for EXCON Personnel

Because this wargame is not exactly like any other wargame, EXCON personnel will need a certain amount of exposure to and practice with this ruleset before they will be able to run a smooth game. Playtesting and repeated IWXs have revealed benefit from familiarity with the flow of the game, experience with the scoresheets, etc. New members of EXCON frequently note that there is a bit of a learning curve!

All EXCON personnel should carefully read this entire ruleset (and the associated *Player's Guide*) as part of their preparation for the wargame. Even personnel who have participated in a previous iteration of this wargame are encouraged to reread the rules. The rules cover a wide range of circumstances and contingencies and so are fairly dense, especially for a first-time reader. Even a careful read is not wholly sufficient to fully prepare EXCON personnel. EXCON should also schedule one or more rehearsals, dry runs, walkthroughs, or "Map EX"–type preplanning sessions to ensure that everything goes smoothly. Such rehearsals might include any or all of the following topics:

- Introduction to the purpose of the game
- Review of the vocabulary and terms of the game (see Section 1.3)
- Introduction to the artifacts of the game: the game boards and pieces, pointers, timers, the scoresheets, the dice
- Assignments of EXCON roles for each step of each turn (see Section 3.3)
- Discussion of the responsibilities in each step for each role
- Guidance for S3 role-players
- Practice for the members of the narration and scenario team who will give the Step 1 update briefing
- Guidance for the narrator
- Practice for the narrator in describing different outcomes (success, partial success, near miss, failure) for an action
- Guidance for the judges
- Practice for the judges in using the scoresheets to score actions
- Discussion between the judges about the different criteria and the judging standards
- Step-by-step pacing and spacing of the game: who needs to be where, and when
- Layouts of the rooms: where different EXCON personnel will sit, especially for Step 4
- Briefings related to the scenario, including the core "storyline" for maneuver
- A chance to ask questions and have them answered.

Members of EXCON should also be fully familiar with the exercise scenario, both whatever materials have been or will be provided to players as background on the operation and whatever materials have been prepared as part of the storylines for maneuver and outcomes as the game progresses (see Section 3.4.5).

3.3. EXCON Personnel and Roles

Table 6 matches activities across the different steps (the rows) with EXCON individuals and their overall role (the columns), listing minimum core personnel and where additional EXCON personnel could be assigned roles. All assignments of roles to specific individuals are provisional and may need to be tailored to the specific personnel and circumstances of the actual game. Table 6 includes some distinctive notation:

- "X" indicates that the activity will normally be performed by the EXCON member listed at the top of the column, if there are enough EXCON personnel for that role to be assigned.
- An asterisk (*) indicates that an activity is suitable for assignment to another EXCON member if necessary or practical.
- "A" (for "alternate") indicates that the activity is suitable for assignment to the EXCON member listed at the top of the column instead of or in addition to the member indicated by the "X" in the row.

TABLE 6

EXCON Personnel, by Role and Activity

	Required Personnel				Optional Roles for Additional Personnel						
Activity	EXCON Lead	BLUE S3	RED S3	Narrator	Notetakers (2)	Action Officer or Stage Manager	Map Keeper	Reality Master	Timekeeper	GREEN Representative	Additional Judges (2)
Manage participant location and timing	A					X					
Update BLUE		A		A			A	X*			
Update RED			A	A			A	X*			
GREEN brief (optional)				A				A		X*	
EXCON representative(s) available to answer questions				A	A	A		X*			
S3 role-player for BLUE		X									
S3 role-player for RED			X								
Run adjudication process	X										
Judging	X	A	A								X
Timekeeping				A	A	A		A	X*		
Game board management				A	A		X*	A			
Notetaking					X						
Narration of action outcomes				X							
Adjust outcomes/storyline from results (team effort)				X							
Deliver turn outcome summary				X							
Consolidate changes to storyline for step 1 of next turn (team effort)				X							

"X" indicates that the activity is typically performed by the EXCON member listed at the top of the column.

* indicates that an activity is suitable for assignment to another EXCON member if necessary or practical.

"A" indicates that the activity is suitable for assignment to the EXCON member listed at the top of the column instead of or in addition to the member indicated by the "X" in the row.

If the S3 role-players are going to be used for an EXCON role outside the control room other than as an S3, then these role-players should have an unmistakable indicator, such as an around-the-neck placard or badge, indicating when they are "on" in their S3 role versus serving as just another member of EXCON. Ideally, and personnel permitting, the role-players should participate as the S3 only in Step 2 and in "control room" activities during other steps, unless they are required to play additional roles. There is no problem with S3 role-players being part of the EXCON team for Step 4, as Step 4 is pretty much all hands on deck; if an S3 is used as part of the judging panel, then S3s from *both* teams should be judges, as it might be perceived as unfair to have one side's S3 as a judge without the other to balance things out.

OPTIONAL: As a further optional role, EXCON might include someone explicitly responsible for representing GREEN. GREEN status is typically maintained as part of the scenario and as a general EXCON duty, and GREEN status updates are typically provided by the member of the narration and scenario team that provides the Step 1 updates, but a separate EXCON role could be assigned for these responsibilities. If there is an EXCON GREEN representative, that individual could also be used on the panel of judges for Step 4.

OPTIONAL: Playtesting and previous IWXs revealed a need to have someone with control over and the final word on reality within the scenario: what works, what doesn't, and the status and condition of things not spelled out explicitly in the scenario. We call this role "reality master," but it could also be called "king of the world" or "scenario control." This role could be filled by the EXCON lead, but playtesting also revealed a relatively heavy burden on the EXCON lead from existing administrative and other duties within the game. This role could instead be filled by a member of the narration and scenario team assigned that task. It is certainly feasible to have the map master/game board manager and the reality master be the same individual. The reality master would explicitly have the following responsibilities:

- Be the final arbiter of any RFIs.
- Be the final arbiter of capabilities available at any time for each team and for facts about their capability and functioning.
- Track all answers given and rulings related to the above points to ensure continuity and consistency in the scenario world.

OPTIONAL: Experience with the game has revealed a heavy load for EXCON lead. One way to lighten that load is to shift some of the EXCON lead's administrative burden to an action officer or stage manager. This individual can take charge of "making the trains run on time" and take care of administrative responsibilities such as maintaining the list of which players will be assigned to which teams, producing and handing out nametags for players, making sure players have the forms and network access they need, letting players know when they should arrive each exercise day, letting players know how long meal breaks should be (and when they are expected to return to the exercise site), letting players know when the next game step will begin and where they should be when it begins, letting judges know what time they need to be ready to listen to action presentations, letting distinguished visitors know what step the game is in and where they can sit, etc. Such an individual should have passing familiarity with the game and its flow, but it is far more important that the action officer or stage manager have good leadership and organization skills and be comfortable "herding cats" and making sure everyone knows where to be and when to be there.

3.4. Preparing the Scenario

This ruleset is designed to be scenario-agnostic, provided that the scenario pits a notional BLUE force against a notional RED force in an operational context where a plan can and should include a concept of information. But the scenario itself requires a great deal of detail and preparation. This section provides guidelines and tips for developing a scenario and providing a sufficient amount of detail within the scenario to meet the requirements of this wargame.

The scenario needs to include sufficient detail to allow teams to plan information activities to support their side's operation. This should include information about context, with a focus on things relevant to the information environment. The scenario should include details on things that may not be included in standard intelligence preparation of the operating environment, such as information on the intentions of various noncombatant groups, the different communications infrastructure present and use patterns, important leaders within the two forces, the personalities and proclivities of those leaders, and characteristics related to will to fight for both RED and BLUE forces. (See Section A.2 for a more comprehensive treatment of will to fight.)

Initial briefings should only include information that is realistically available to the two teams in their respective roles, but other information related to the information environment that might be uncovered through RFIs or through experience should be included within the full scenario details so that the narration and scenario team is not called on to invent such details on the fly during the game. The best way to ensure that a scenario is sufficiently comprehensive to answer a variety of questions that may arise is through playtesting.

Remember that this is a two-sided game, and full intelligence preparation of the operating environment materials must be provided to both sides.

Scenario information provided to each side must also include lists of the various maneuver elements involved in their force, the information capabilities that are organic to their force, and information capabilities that might be available by request from a different echelon or adjacent force. Each side should receive that same information about the opposed force, filtered or obscured as appropriate because of realistic intelligence constraints.

In many wargames, the players or participants are the source of all plans for the formation they represent, and the scenario only supplies the starting position and context. This is not the case for scenarios for this wargame. Because the players only develop and contribute the concept of information, all other parts of the plan (including the concept of maneuver and the concept of fires) and activities by their notional side should be part of the scenario. So, the scenario is not just the starting position and context but is also the story of exactly how the whole operation will unfold absent the input of the teams' information activities.

NOTE: The results of information activities may cause EXCON to need to slightly adjust the story, but the story of the operation written as part of scenario development will pretty much stand (see Section 3.4.5). One EXCON playtester described the situation like this: "Imagine that the operation, the game, is over. Now write the history of everything that happened except the information environment part. That is what we need to know for the scenario."

Obviously, only the starting position, context, and initial maneuver plans for their side are provided to players. But the planned evolution of the context and the plans for both sides, as well as the expected maneuver outcomes, need to be written as part of the scenario materials available to EXCON. Scenario developers might think of the task as developing two sets of materials: One set is the standard background, context, and mission planning information required for any scenario-based exercise or wargame, and a second set provides additional scenario details, including not only important facts that aren't in the materials provided to the players, but also the story of how the operation will unfold and all of the details of what will happen to maneuver forces later in the game. See Section 3.4.5 for further details.

3.4.1. Required Level of Detail in Full Formation Plans for BLUE and RED

The two teams represent just the information working group (or equivalent) in the staff. All remaining staff sections are controlled by EXCON and represented to the players as abstractions or in the persons of the S3 role-players. This means that part of scenario preparation and maintenance includes the creation of all aspects of the RED and BLUE plans that are not related to the concept of information. That includes the base plan that the concept of information will support and the evolution of that plan as it unfolds during the game.

Maneuver plans from other staff sections built into the scenario need to be sufficiently detailed to allow the teams to plan a concept of information to support them, and maneuver plans need to be sufficiently detailed for the narrator to describe what happens in the operation at the end of each turn. An updated maneuver plan will need to be provided to each team at the beginning of each turn. These should be part of the pre-prepared materials that are part of the scenario.

Playtesting revealed the importance of having fairly elaborate and detailed plans for maneuver elements. When maneuver objectives and intended actions were vague, playtest players struggled to connect their proposed actions to aspects of the mission they were supposed to support. Because connection to operational objectives is one of the scoring criteria for evaluation and adjudication of actions, clear and detailed objectives and maneuver plans are required in order to be fair to the players.

3.4.1.1. Overall Mission Objectives and How They Relate to Teams' Concepts of Information

The detail in broader staff operational plans should include clear overall objectives that the teams' concepts of information can then support. When players' objectives for their concepts of information or their specific actions have been deficient or failed to nest clearly with higher-level objectives, the fault always appears to have been with the level of detail and clarity provided in the overall objectives. This has particularly been the case when IWX scenarios have not been combat scenarios but instead been competition-focused (see Section 5.2).

There are a number of steps that might be taken within the IWX Wargame to improve objectives. There are two general approaches possible. First, revise the scenario-provided guidance to make it easier for players to connect good objectives. Second, guide and instruct players during initial planning and Step 2 planning to improve objectives.

One way in which scenarios might be adjusted to promote better player-determined objectives is to include higher-level guidance or objectives that are more specific and easier to nest with or connect to. There is a temptation, especially when conducting a classified wargame in a realistic scenario context, to simply use existing real regional guidance and objectives. This has been done in several actual IWXs, and it has the advantages of requiring little work (the objectives and guidance already exist) and having good realism (they are the real guidance and objectives), as well as being something with which players should already be familiar or something that the exercise can help reinforce.

However, even though these objectives are realistic, they might not make for a good wargame. If theater objectives have a theater timescale, it might be very hard to imagine a meaningful increment of progress over the exercise timescale. Good guidance for an IWX Wargame (be it at the tactical or the operational level) will specify objectives that can be met or meaningfully improved upon during the exercise timescale that players' supporting objectives can then contribute to.

Scenario designers do not need to jettison the existing higher-level guidance or objectives, they just need to develop some exercise operation–specific intermediate-level objectives. For example, unpublished RAND research has identified three canonical purposes for security cooperation: building partner capacity, gaining or maintaining access and willingness, and expanding interoperability. Higher-level guidance related to competition within a specific region is likely to include variations of one or more of those themes. Given the focus of IWX on influence, "access and willingness" are likely to be points of emphasis within an operational-level competition-focused game (see Section 5.2). If existing guidance mentions a desire to increase access or partner willingness, then scenario operation–specific guidance might specify a discrete step in access or willingness that the operation aspires to achieve, such as persuading certain key leaders in a partner nation to agree to a follow-on exercise or event, or getting a partner to demonstrate willingness by committing new forces to a future multinational exercise.

Scenario objectives can also be made easier to connect to by specifying clear and specific objectives for the maneuver portion of the supported effort or for the overall mission. Perhaps the RED force mission involves a hospital ship and a humanitarian mission, with some small increase in goodwill as one of the competition-type mission objectives, which will likely lead players toward safe and modest objectives. That RED force could also be given a clandestine mission, perhaps to use the presence of the forces with the ship as an opportunity to coerce or bribe a specific partner nation politician, or to stir up local animosity against the BLUE force. While still a competition-type objective (intended to stay below the threshold of conflict), the example clandestine mission will push players toward more aggressive gray zone actions that are closer to the threshold of conflict, and to actions that are riskier because of the increased risk of either failing to meet the more immediate objective or failing to adequately support a clandestine activity through operations in the information environment (OIE).

In addition to providing players with more provocative and engaging intermediate or mission-specific objectives, players can also be instructed to propose stronger objectives and guided on how to do so. This can be a point of emphasis during exercise instructional blocks, a key area of focus for teams' mentors, and something that approvers (either the S3 role-players or the recipients of a confirmation brief) push the players to improve.

3.4.2. Scenario and Plan Information Required for Adequate Preparation of the Role-Players and Adjudicators

The game requires an S3 role-player for each team. To allow these role-players to function effectively, they need to be familiar with the details of the context, the details of the operation's progress to date (prior to each turn), and the intended progress of their side for the turn. **The role-players need to know everything the players need to know, but they also need to know more.**

Specifically, the role-players need to know the answers to the kinds of additional questions that the players will have so that the role-players can preserve credibility in their role by knowing more than the players. Some of this can be achieved by the role-players reading the scenario introductory material more carefully than the participants, but some of the information in the scenario should initially be EXCON-only material—things that players might (or should) ask as RFIs. Such material might include current specific locations of friendly or enemy formations (especially information capabilities), information about the personalities of formation leaders (on both sides), information about the status or intentions of GREEN groups, expected weather, etc.

An example of the sort of material that might be useful for EXCON to include is the data tracked in Figure 10. Figure 10 depicts the summary assessment of the will to fight of the two scenario forces at the outset of the scenario broken down into component elements.[7] Tools like this could be used to help EXCON track important changes in the game.

Game aids such as a will-to-fight tracker could also be displayed to the players to give them a better idea about where actions might be successful and the evolving influence landscape they face. If such summary tools are displayed to players, we recommend several caveats. First, it should be made clear to the players that these scoresheets do not represent any intelligence or staff product that actually exists, and that during a real operation they would not have access to a definitive assessment of these factors and would likely have to assemble their own best guesstimate. Second, such tools should be displayed only in the engagement room and should be noted to be (at least partially) covered by the "what happens in the engagement room stays in the engagement room" rule described in Section 2.5.4.2.

EXCON also needs to have fairly comprehensive tables of organization and equipment for both sides, not just for the information capabilities available to each but also for the other capabilities, especially various different systems and platforms. Because the inherent informational aspects of all military activities can be leveraged as part of a concept of information, what capabilities are available is an important part of the scenario baseline. In addition, **EXCON needs detailed information about the performance and actual capability of available capabilities.**

During playtesting, creative players "made up" the operational effects possible from capabilities they were unfamiliar with; EXCON needs to be prepared to impose the real-world constraints that are inherent in available capabilities. (This could be the responsibility of an individual charged with the optional EXCON role of reality master; see Section 3.3). Ideally, sufficient expertise or pre-study of the capability by EXCON (perhaps by the reality master or the narration and scenario team) will enable EXCON to provide needed answers when questions of capability come up.

[7] For more on will to fight, see Ben Connable, Michael J. McNerney, William Marcellino, Aaron Frank, Henry Hargrove, Marek N. Posard, S. Rebecca Zimmerman, Natasha Lander, Jasen J. Castillo, and James Sladden, *Will to Fight: Analyzing, Modeling, and Simulating the Will to Fight of Military Units*, RAND Corporation, RR-2341-A, 2018.

FIGURE 10

Will-to-Fight Summary Comparison of the Initial Status of the Two Opposed Forces in IWX 20.2

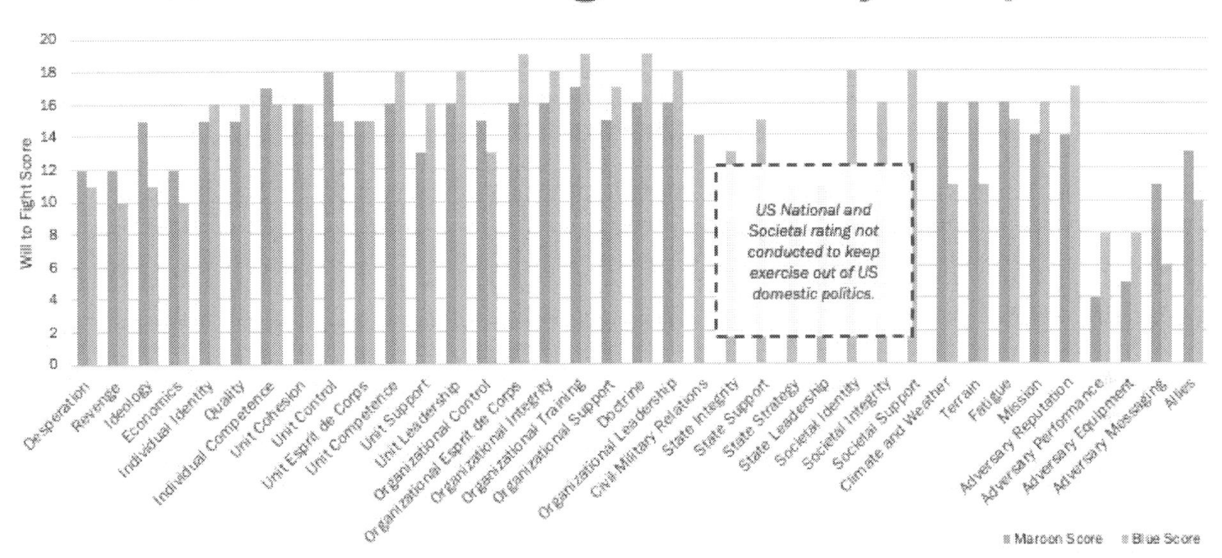

If the answers aren't known, it creates a learning opportunity for both players and EXCON. Also, see Section 2.6—the show must go on. If getting to adequate information on the real-world functional parameters of a capability is going to take too long and disrupt the flow of the game, EXCON should impose an assumption (either their best expert judgment or something based on a die roll) so that the game can continue. (EXCON should also do their level best to track down the correct answer and share it with all participants as part of the game's AAR to facilitate learning.)

3.4.3. Scenario Details Related to the Game Boards/Maps

This game is intended to be played with an associated scenario-specific set of game boards (which might be actual boards or might just be large maps printed with a plotter and laminated). The content of these boards or maps needs to be provided as part of the scenario. The scale and size of each map should be determined based on the sizes of the two maneuver formations and the expected scope of the operation (both in terms of distance and of time). This requires some thought to make sure that the operation uses or unfolds across much of the board/map (it would be a shame to have the whole operation unfold within 4 square inches in the middle of a table-sized map) without having too much or any of the relevant action taking place "off map." (It is okay if some capabilities fly in from off map, or if naval fires are directed from off map, but, ideally, all of the actions, their target audiences, and their effects can be portrayed on the game board.)

In addition to the actual map or board, the scenario needs to include some details about the physical space represented by the map: for example, the length of a key route and an estimate of how long it will take forces to transit it, or an estimate of throughput of local traffic. Information about the physical characteristics of the information environment could also be provided: the broadcast footprint of a TV station, the locations of cell towers, etc. Such information might be stored as a "layer" in a digital version of the scenario map that could be referenced by narration and scenario team briefers and then used to pinpoint (and perhaps indicate with a token, marker, or grease pencil) some key feature of the physical information environment.

The actual representation of the game board and markers or tokens is discussed further in Section 3.8.

3.4.4. Level of Detail Regarding GREEN Required in the Scenario

As has been mentioned, the scenario often needs to include details about culture, context, and local groups, often commonly referred to as GREEN (to distinguish them from the primary opponents, BLUE and RED). Because this is primarily an influence game, one or more groups under the purview of GREEN are likely to become a TA for actions by one or both teams. To support this, the players will need information about such groups, so such information should either be part of their intelligence briefing or something they can request as an RFI.

Details should include the type or composition of different groups or population segments, baseline behavioral proclivities, relevant cultural characteristics, significant narratives, views on RED and BLUE, relationships with other groups, leaders/influencers, media use patterns, etc. The more specific the information provided in the scenario, the better the teams will be able to tailor precise and effective actions (and presentation of those actions) to match, and the better the judges will be able to assess the prospects for success of such actions.

3.4.5. Scripting a Range of Operational Outcomes Possible at the Conclusion of Each Turn

As noted in Section 3.4, the scenario should include the story of how the whole operation will unfold except for the information environment parts. This is referred to elsewhere in the ruleset as the *storyline*. The initial scenario materials available to EXCON should have a likely storyline that covers the expected flow of the operation across all five turns of the wargame. The storyline needs to be somewhat flexible in case the teams' actions have a significant impact on progress toward one or more of the operational objectives. Even leaving room for that flexibility, the scenario should provide prepackaged materials for the Step 1 update briefing for every turn that can be slightly adjusted or tailored based on any dramatic effects from information activities in the previous turn or turns.

As noted in Section 2.6.1, the wargame has a range of possible outcomes for each turn, driven primarily by preplanned maneuver outcomes but capable of being affected by information activities along three tracks or storylines: (1) an expected direction of progress, (2) progress based on RED information activities outperforming, and (3) progress based on BLUE information activities outperforming. If teams and dice are relatively balanced, turn-by-turn progress of the overall operation should remain within these left and right bounds.

Depending on the time available for scenario development, the scenario should include a fully developed storyline and set of update briefings for track 1, the expected unfolding of the operation. Tracks 2 and 3 may be more or less developed. Ideally, all three storylines will be fully developed and the narration and scenario team can decide between Step 5 and the following Step 1 which of the three storylines the game is closest to and then quickly customize or tailor from there based on actions and their outcomes. With three fully developed storylines, it is relatively easy to add a few details to discriminate more subtle variations. Care must be taken to keep things from getting out of bounds! If one team (say, BLUE) has repeatedly outperformed and held the advantage, don't let the storyline get any better for them than the most favorable storyline you've developed for them. Describe things as continuing to go well for them, but use the "things are going well for BLUE" storyline. And consider making things harder for BLUE and steering back toward the core storyline by adding a corrective inject (see Section 2.6.1.2).

See Figure 11 for an example of possible outcomes and keeping them under three core storylines.

The overall level of detail in the storyline should correspond exactly to the level of detail needed to successfully complete Step 1 for each turn after turn 1. That is, there should be a prepared briefing covering all Step 1 topics waiting to be very slightly tailored based on the previous turn's Step 5 and presented to the teams so they can plan their next actions.

FIGURE 11

Notional Depiction of the Storyline and Its Possible Deviations Turn by Turn

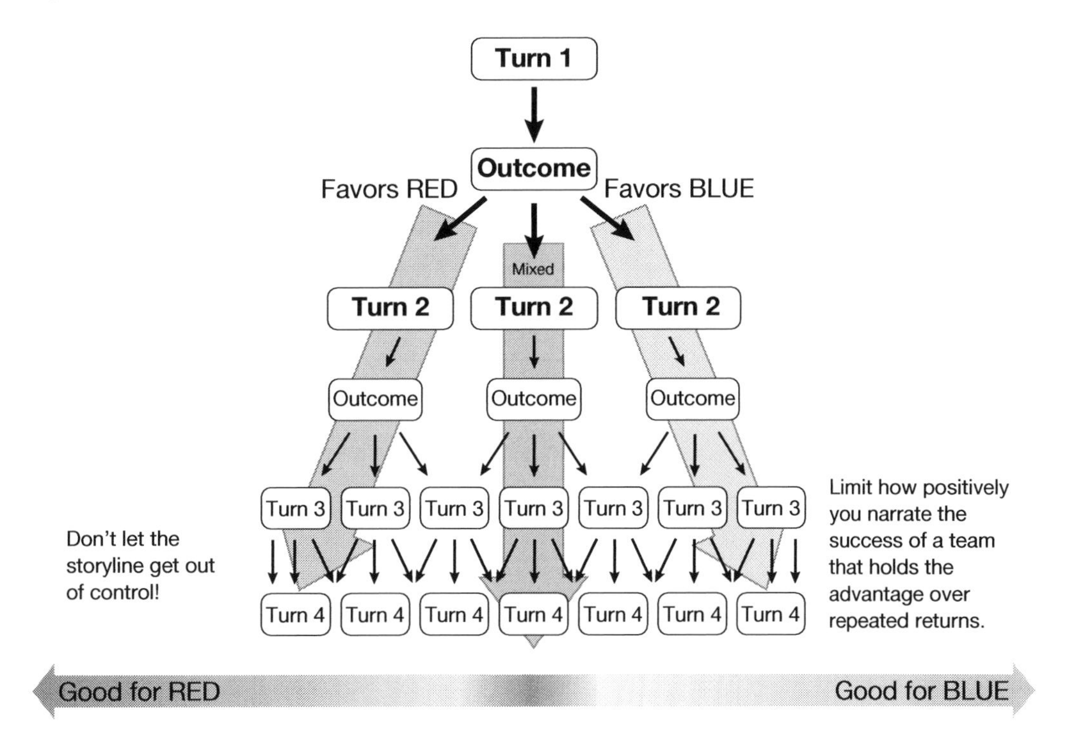

To develop the left and right bounding storylines, scenario designers should imagine possible effects that information activities might have on maneuver and on the overall progress of the operation. It is difficult to imagine exactly what the teams will prepare and impossible to anticipate which actions will be accompanied by sufficient dice luck to succeed. However, the range of possible effects and their impact on maneuver is easier to imagine. Many different information environment effects might keep an enemy formation from moving: interrupted or falsified command networks, diminished will to fight, suppression, causing noncombatants to block routes, causing enemy forces to fear movement, causing enemy commanders to believe it is in their interest to remain in position, etc. There are many possible ways to have the same effect on the game—namely, that a formation doesn't move or doesn't move as far as intended.

Opposite to actions that have the effect of slowing the progress of a formation (or a whole side) are actions that might speed the movement of a formation toward its goal. Actions that remove obstacles (getting a blocking formation to move to a different location or retreat) or reduce resistance (a feint that causes a formation to orient in the wrong direction, making it vulnerable to a flank attack, or successful efforts to keep refugees from clogging roads) could similarly speed progress. The basic outlines of left and right bounds for storylines, then, should probably focus on one set of possible outcomes in which the progress of the attacking force (if the scenario contains such) is slowed or delayed, and another set in which the progress of the attacking force is easier than anticipated and accelerated. General outcomes along these lines could then be tailored to be described as a consequence of whatever the prominent effects were from the game actions of the previous turn.

NOTE: Many of the effects of information need not actually change the pre-scripted scenario storyline: They can be *described* as having had the intended effects on the TAs and having enabled operational objectives without requiring the planned storyline progress to change. This is *not* to suggest that information activities do not have important effects that impact operations and campaigns! Quite the contrary; the authors are strong believers in the importance of effects in and through the information environment. However, just because information activities have effects doesn't mean they *actually* have to change the pre-scripted flow of what happens across turns in the game: Because the players don't know the pre-scripted storyline, EXCON can *describe* outcomes and progress as changing, when in fact they are unfolding according to script. The narrator can keep the game exciting by describing the effects of an information action on the game, but the extent to which that action actually affects the outcome of the game depends on their best judgment and the overall flow of the storyline.

Between this opportunity to narrate effects without actually significantly changing preprogrammed progress on the storyline and left- and right-bounded alternative storylines, it should be possible to keep the game within these anticipated tolerances. Should one team's actions be so consequential as to start to push the storylines out of bounds, perhaps it is time for a corrective inject. See Sections 2.6.1 and 2.6.1.2.

3.4.5.1. Example of Storyline in Practice

To make the notion of the bounded storyline depicted in Figure 11 more concrete, this subsection walks through an explicit example.

Figure 12 depicts the progress of this example of deviations within the storyline. As discussed in Section 3.4.5, EXCON (specifically the scenario and narration team) for this notional example has prepared three general storylines for progress through the five turns of the IWX Wargame: one that favors the RED team, one that represents a mixed outcome for both RED and BLUE and is based on what would happen if the two competing information planning cells basically neutralize each other, and one that favors the BLUE team.

In this notional example, RED has two of their three actions succeed and produce important effects in turn 1, while BLUE has none of their actions succeed. The narration and scenario team decides that the outcome has been pushed strongly into the "favors RED" storyline, and that is what they will use for their narration in Step 5 of turn 1 as they describe what has happened (with the description adjusted to match to the specific actions undertaken by RED). In Step 1 of turn 2, the scenario and nar-

FIGURE 12
Example of Turn-by-Turn Storyline Deviations

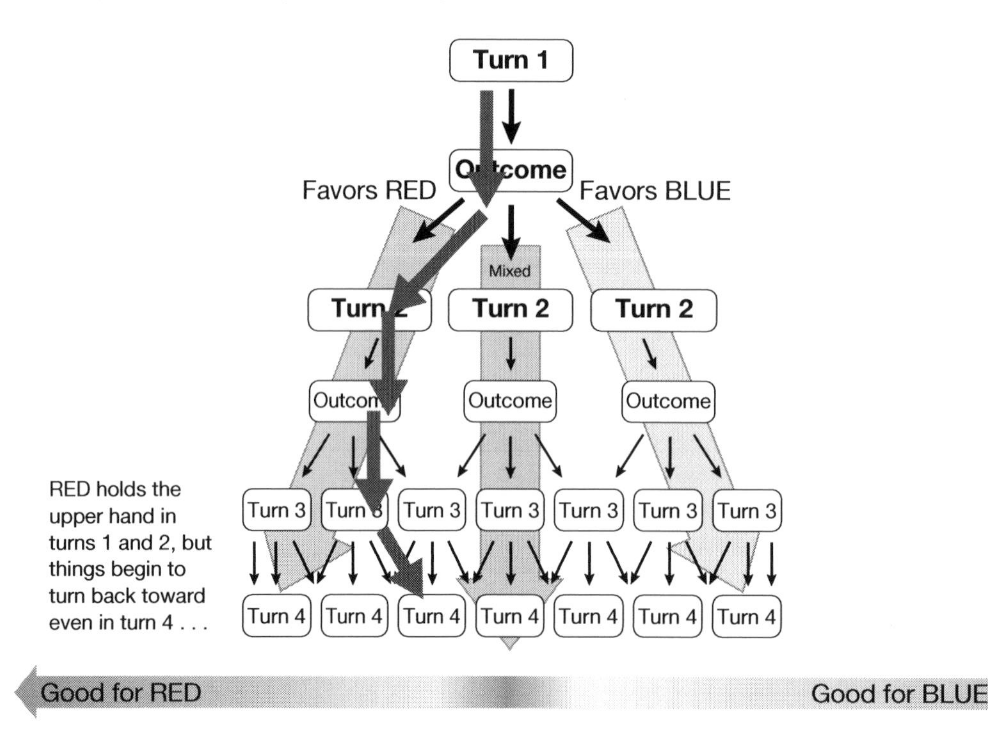

ration team will present the update to both teams based on the favors RED storyline.

In turn 2, the outcome is more mixed, with both RED and BLUE succeeding at one action each, and other actions failing to execute. This mixed outcome for the turn still keeps the storyline in the "Favors RED" lane; the Step 5 narration at the end of turn 2 will note the trading of information blows between BLUE and RED but will still describe conditions that describe RED making better progress toward its overall objectives than BLUE, and will use the "Favors RED" storyline conditions for the start of turn 3. In turn 3, RED has one action that is slightly successful, while BLUE has two actions succeed, one of which has effects that are fairly important. This steers the storyline back in the direction of BLUE. For the Step 5 narration in turn 3 and Step 1 situation update in turn 4, the narration and scenario team will use the "Mixed" storyline, with a tinge of RED advantage, to reflect that conditions are now pretty close to the main mixed storyline, but with RED having enjoyed slightly more success overall than BLUE.

Recall that RED's success had pushed the storyline over to "Favors RED" right from the first turn. Had RED achieved another strong turn relative to BLUE in turn 2, they might have been pushing the bounds of the "Favors RED" storyline for turn 3 (that is, threatening to push out of bounds toward a story more favorable to RED than the "Favors RED" track). The scenario and narration team might handle that in several ways (all discussed elsewhere in the rules):

- As per Section 2.6.1.2, the narration and scenario team might have used a corrective inject to make things harder for RED and make the storyline more likely to push back toward the middle.
- As per Section 2.6.1.1, the narration and scenario team might have elected to have turn 3 be a repeat of the initial conditions from turn 2, with BLUE making no progress toward their objectives (assuming that RED was on the defense), or fast-forwarding conditions to conditions for turn 4, effectively skipping turn 3, assuming that RED's highly successful efforts had accelerated progress toward RED's objectives (assuming RED is on the attack).
- Or, as per the whole idea of bounded storylines (Section 2.6.1), the narration and scenario team might simply keep the storyline stuck at "favors RED," narrating the strongest version of that storyline, but not allowing the overall story to escape the established left and right limits.

NOTE: The storyline favoring one side or the other just means that things are going well for that side in terms of the progress of the pre-scripted maneuver aspects of their side's overall plans—that the things that team has done have worked enough to help support their overall effort. It does not necessarily mean that one team has won durable advantages in the information environment. Durable advantages should come from tracking effects on specific TAs across turns, or from actions that build on cumulative effects.

3.4.6. Preparing Scenario Injects

Section 2.6.1.2 provides a discussion of using injects in the game for three purposes: to stimulate players to more interesting actions, to challenge teams with unanticipated problems, and to increase difficulty for one team to correct the overall progress of the storyline and get it back within tolerances. Section 2.6.1.2 suggests that injects be presented as cards to both teams to create a dynamic feel within the game and give players something concrete to respond to.

Injects are not strictly necessary. Injects have been used sparingly in playtesting and in IWXs to date, and the games have all been sufficiently dynamic with just the chaos caused by the two teams and their actions. Injects are optional.

Although injects should *feel* dynamic, they need not be. Injects *can* be changes to what was originally planned, but they can also be preprogrammed parts of the storyline that are *presented* as if they were dynamic. If the storyline includes some preplanned setback or misfortune for the maneuver progress of one side, we recommend that that be presented as if it were actually a dynamic inject!

Remember when designing injects that the goal is to stimulate or challenge the players, *not* to make things more difficult for EXCON. Injects that are likely to cause big swings in the storyline may seem like a good idea but are likely to be unnecessary and will make the game harder to run.

If injects are going to be available to be used, a set of injects should be prepared as possible inclusions with the storyline update materials for Step 1 of each turn. They should be cards that can be given to each team. Because each side may perceive the events of an inject differently, cards may need to be team-specific. For example, a RED unit leader being executed when his infidelity with his commander's wife is exposed might have different indicators for BLUE or for RED, and, when it is reported in the media, the two sides might assign different levels of credibility to the report.

For injects to be interesting, they need to either create an opportunity or a problem within the information environment. Also, injects should be written and described as if they are chance events. Injects should *not* be information activities from elsewhere in the force (a higher echelon, an adjacent formation). Injects should appear spontaneous and accidental. Some possible examples of injects:

- A key influencer is accidentally killed by the operations of one of the two sides.
- A key leader of part of one of the subordinate formations is killed in combat, or is relieved with cause, and either leaves that formation rudderless (and perhaps more vulnerable) or is replaced by a new leader who has different proclivities.
- International outrage descends on one side because of reports of atrocities (whether these accusations are true and who made them might be pointed RFIs).
- Something about the operation or the response leads to protest in the area (peaceful or otherwise).
- A BLUE helicopter goes down, and recovery operations (and recovery support operations) become necessary.
- The presence of noncombatants in the area is much higher than intelligence suggested because of a new local shrine (or an unanticipated local holiday or festival, or some other reason), and this increased presence is a threat to operational progress for one or both sides.
- A BLUE vehicle hits and kills a child; it is unclear what the response throughout GREEN will be.
- A commander or leader on one side is embroiled in a personal scandal that breaks in the news today.
- Rolling power outages restrict internet and mobile phone availability across part of the area of responsibility (even where batteries are charged, the towers are dark).

Injects need to either include or be backed up by sufficient detail. Think about the kinds of RFIs that an inject is likely to generate, what the actual answers are, and what each side's intelligence capabilities will be able to discern. Preparing this via playtesting is a good way to ensure that you have the right information ready at the start of the game.

Injects require sufficient detail, but not an extensive amount. They need to provide sufficient initial information to stimulate a response from the players, and the narration and scenario team should develop a little more information about each inject they might use, including additional details that might be available to players as events unfold or if they make an RFI, and some idea of what might happen if neither side does anything in response to the inject (how it might affect the storyline or the information environment). So, for example, an inject might be "There are rolling power outages across Bender City that are adversely impacting internet and mobile phone availability." Additional information available on investigation might note the specific areas affected by the outage (could be drawn on the enquiring team's map), that a specific substation (which could also be located on the map) has been damaged in the fighting, and that the relays necessary to draw power from elsewhere in the city are not in place. Additional planning notes could include a GREEN note regarding which side the relevant population blames for the outage, and the fact that internet or mobile phone–based or –reliant messaging is much less likely to be able to reach intended audiences in the affected area until repaired. Inject notes might also say how long it will take indigenous civil elements to repair the outage, and whether that could be hastened through support or encouragement from one side or the other.

3.4.7. Time Represented by Each Game Turn

Section 1.3.2 notes that the amount of time represented by a turn within the wargame may vary throughout the game, and further notes that turns in which the tempo of maneuver forces is relatively low may represent longer periods of time. Section 2.1.2 notes that the update briefing given in Step 1 of each turn will inform players of the amount of time the coming turn will represent. Intended time to be represented by each turn should also be anticipated and specified as part of scenario design, and confirmed as the narration and scenario team matches the scenario to the specific exercise. What considerations should affect EXCON decisions about how much time to make each turn represent?

Several factors matter in this decision. One is the desired span of time within the scenario that the wargame is intended to cover. This, combined with the total number of game turns intended, can give EXCON an initial rough idea of how much time each game turn should represent. However, all turns need not represent equal amounts of time. It would be perfectly reasonable to have a game in which turn 1 represents a week of shaping and coordinating activities while marines are underway, and turns 2–4 each represent a single day of action as marines come ashore and begin their operation, with turn 5 being either another single day of operations or a longer period (three days? a week?) covering transition or consolidation efforts after the core of the operation, depending on the progress along the storyline toward completion of the operation.

Other considerations include the pace of operations and the realistic pace of the staff section that players represent. During high-intensity phases of operations, such as combat, turns might represent less than a day. However, it would probably be unrealistic to have turns represent any period shorter than the actual time of play for a turn. Turns might represent longer periods (several days) or might represent a day but have a gap of several days narrated between turns.

A competition scenario rather than a conflict scenario poses additional challenges related to time represented. Conflict unfolds at high tempo and is very dynamic, often including short cycles of action and counteraction both in the information environment and in the spatial domains. Competition generally unfolds much more slowly and across a longer time span. However, as risk increases and activities come closer to the threshold of conflict, or during crises, competition activities can unfold very quickly. As an example, the IWX 21.1 scenario (a competition scenario) covered a span of about two months. Turns represented a little over a week of time, focused more on port calls and specific locations in sequence rather than precise and equal blocks of time. This worked reasonably well for that scenario. One could also imagine a competition scenario game with more uneven time representation: perhaps a turn (or two turns) of much longer term shaping or preparation efforts (a month or more each), and then short turns surrounding the core events of the game.

As general advice, EXCON should plan time represented per turn in advance and in close consideration of the overall storyline of the scenario (the way in which the scenario is expected to unfold—see Section 3.4.5). Within that storyline, where are the key junctures or milestones at which teams should seek to have an impact? Ideally, each of these key junctures will take place during a single playable turn, and the time represented by each turn should be apportioned so that turns align with these key scenario junctures. In a conflict scenario, these junctures might well be combat progress milestones, and success for one side might be a delay in the other side's progress toward the next milestone, requiring another turn of actions at the previous milestone. In a competition scenario such as IWX 21.1, each key location to be visited by the supported formation might be a key juncture or milestone, with efforts prior to departure being another possible key juncture. A crisis (perhaps a planned inject?) at one of the key locations might necessitate an extra turn at that location as the two teams take actions and counteractions more rapidly based on the crisis and the other team's response.

NOTE: The rules strictly forbid counteractions within the counterarguments and rebuttals (see Section 2.4.3.4) unless these counteractions come from approved battle drills or standing operating procedures (see Section 2.2.5.1). Because of this, when one side has the opportunity to take actions that are likely to demand a response, it is a good idea to have the time represented by the turns be short enough that actions in response taken in the next turn still feel like response actions, rather than being hopelessly late or overtaken by events.

3.5. Preparing Academic Instruction and Preliminary Planning

Playtesting has revealed that an academic period (for cross-leveling and instruction) and a planning period prior to the start of the wargame are essential to success.

At minimum, the academic period must include instruction regarding the terms that will be used and establishing a shared baseline of understanding about information capabilities and planning for participants to be able to meet expectations regarding plans for actions to be proposed.

Similarly, the time allocated for Step 2 in the flow of each turn of the wargame is not sufficient for participants to digest the scenario and plan a concept of information. Such planning (and related familiarity with the scenario and maneuver plan) should be begun before the wargame begins. The better the quality of the teams' existing plans prior to the wargame, the better (and more smoothly) the game will go.

3.6. Estimates of Timing for Play

As noted, turns consume a full day of game play, either one exercise day, or an afternoon and then a morning half-day pair. Table 7 summarizes activities by step and estimates the time consumed by each step historically in playtesting and in actual game circumstances.

3.6.1. Considerations or Changes That Might Require Additional Time

The first time an EXCON group conducts an IWX Wargame, it will likely take longer than anticipated simply because of the EXCON members' unfamiliarity with the workflow of the game and of where players or activities can be expedited and where additional time needs to be allowed. In such a case, expand time allowed for each step and allow that a single turn will take a long exercise day rather than just an exercise day. Similarly, the first time a veteran EXCON team conducts a different style of IWX Wargame (one at the operational level instead of at the tactical level, or with a competition-focused scenario, or with multiple staff layers all participating on one or both sides), things will take longer than anticipated. Plan accordingly.

If initial planning time for players is cut short, then additional planning time will need to be allo-

TABLE 7

Estimates of Time Required Per Step, in Minutes

Step	Activity	Planned Time Limit	Lower Time Estimate	Maximum Time Estimate	Estimate Range
1	Receive situation and update	15	10	30	10–30
2	Prepare to present	90	60	120	60–120
3	Present actions	45	0[a]	60	0–90
3a	Final revisions to actions	15	0[a]	30	
3b	EXCON prep for Step 4	Simultaneous			
4	Engagement meeting	210	180	240	180–240
5	Results and reset	30	15	40	15–40
Total		405			265–520

[a] Under some optional rules (see Sections 2.3.7 and 2.3.8), Step 3 is either omitted or conducted in stride with Step 2 (with no additional allocation of time).

cated to Step 2. Initial planning time might be reduced if IWX itself is conducted over fewer than the standard ten exercise days, or even on a ten-exercise-day model if more of the first week of IWX is spent on academics and training than on planning.

Playtesting revealed that the untimed discussion prompted by EXCON questions, which takes place during Step 4 after the timed presentation, counterarguments, and rebuttal of the two teams, is very valuable in terms of exchange of information and learning. Where possible, extra time should be allowed for this activity. The time estimate of 30 minutes per action includes time for judge questions and responses, time for judges to complete scoresheets, and time for outcome rolls to be made, in addition to the time required for presentation, rebuttal, and refutation related to the action.

3.7. Preparing the Physical Space

This game can be run in a single physical space or separated over virtual space.[8] In both cases, at least three isolated rooms will be required:

- BLUE side planning room
- RED side planning room
- central engagement room.

These three rooms can be collocated in a single building or workspace or separated across a video teleconferencing system. This section describes both approaches. Whichever option is selected, the game also requires a fixed set of gaming materials, described in the next section.

Many layouts for the planning rooms are acceptable, and a flexible information OPT should be able to manage in just about any space. The ideal planning room will have multiple networked computers, chairs for all team members, a table or desks, projection capabilities, and whiteboards or easels.

The layout required for the engagement room is a little less flexible. The engagement room should, at minimum, accommodate the largest game board or map, have places for the two teams to congregate, and have a table where the judges can sit as a panel and receive the presentations. Playtest and experience have revealed an "in the arena" layout as particularly effective.

For the "in the arena" layout, tables are arranged in a rectangle with open space in the middle and gaps at the ends of the rows of tables to allow easy ingress/egress to the central arena area. A large laminated plotter map gameboard is laid on the floor in the middle of the square (see Section 3.8), and both the presenting player and the opposed rebutting player stand on the map "in the arena" for the duration of their pair of actions (the two players swap roles after the first action). Only the presenter and rebutter are "in the arena"—all other players, participants, and EXCON personnel remain along the outside edges of the square of tables.

In the arena layout, chairs line the tables along the outside of the rectangle. One of the long sides is for the judges, the GREEN representatives (if used), and the timekeeper. Opposite the judges sit the rest of narration and scenario team and any note takers. The short edges of the rectangle are allocated to the two teams, one on each edge, with players sitting or standing in a crowd behind their respective rows of tables. Observers and distinguished visitors can sit along the EXCON edges as space permits, sit in "back row" chairs along the edge of them room, or stand behind EXCON or the participants. Figure 13 depicts an "in the arena" layout.

Ideally, the engagement room also includes capabilities to display or project from computers. Monitors or projectors might be used to support player presentations, or updates regarding the state of play, or be used to display time countdowns (a projected countdown timer works just as well as a standalone timer), or the outcomes of actions taken in the game. In lieu or in addition to digital displays, whiteboards, chalkboards, butcher paper easels, or sticky notes can also be used to display information on the walls in the engagement room.

[8] Restrictions imposed by the 2020 COVID-19 pandemic required virtual, remote components in the design considerations from the ground up for the IWX Wargame design. This game could be played virtually, or in a combined virtual-physical design that relies on video projection capability, the relaying of game maps to virtual participants, and the possibility of employing virtual gaming services. Any effort to shift some or all aspects of this game to a virtual platform would require considerable design and rehearsal on the front end, as well as redundant communications to ensure that no single point of communications failure interrupts the game flow.

FIGURE 13

The "In the Arena" Engagement Room Layout Used in IWX 23.3

Photo credit: Jonathan Welch, RAND.

3.8. Preparing the Game Board and Various Representations

This section discusses the various physical (or virtual) artifacts required to play the game.

Game boards/maps: This is a tabletop game played on a flat map surface laid out on either a large table or the floor. As designed, the game board consists of a map with military grid reference system (MGRS) markings at a scale appropriate to the scenario. A larger version of the map is placed in the central engagement room, and a smaller copy of the map is provided to each of the two opposing teams. Apart from the difference in size, the three maps must be identical. For the October 2020 IWX version of the wargame, MCIOC EXCON employed one map that was 172″ × 148″ at 1:25,000 tactical scale in the engagement room (shown in Figure 14), and a smaller map in each of the planning rooms. As of late 2023, MCIOC retains copies of these maps and the map data.

The specific sizes and scales of the maps can be adjusted to fit any selected scenario as long as they are identical to each other and have sufficient space to allow for planning, maneuver, and free play. Figure 15 shows the map used for the IWX 20.2 scenario.

Electronic map alternative: If the game is played entirely in the virtual realm, or replicated virtually, there are several options available for electronic game board depiction. None of these were tested during the IWX 2020 wargame playtest and development period, but they could be explored and tested for future iterations. Options include but are not limited to the use of the Office of the Secretary of Defense Standard Wargame Integration Facilitation Toolkit (SWIFT); the open-source VASSAL virtual gaming system; or Google Maps, Google Earth Pro, or a similar shared mapping service. SWIFT has the capability to host classified gaming.

Physical tokens and icons: Units, population elements, and assets are represented by physical tokens, which are distinguished by their color and the icons on them. The IWX 2020 wargame used a common set of 2″-high plastic tokens set in 1″ round bases. Tokens and icons are used to help players and EXCON visualize the position of units and assets on the map, and also to help track the availability of assets not yet in play. Red tokens (alternatively called game pieces or markers) represent the RED

FIGURE 14
Map Displayed in the Engagement Room During IWX 20.2

Photo credit: Nathan Rosenblatt, RAND.

FIGURE 15
Map Displayed in the Planning Room During IWX 20.2

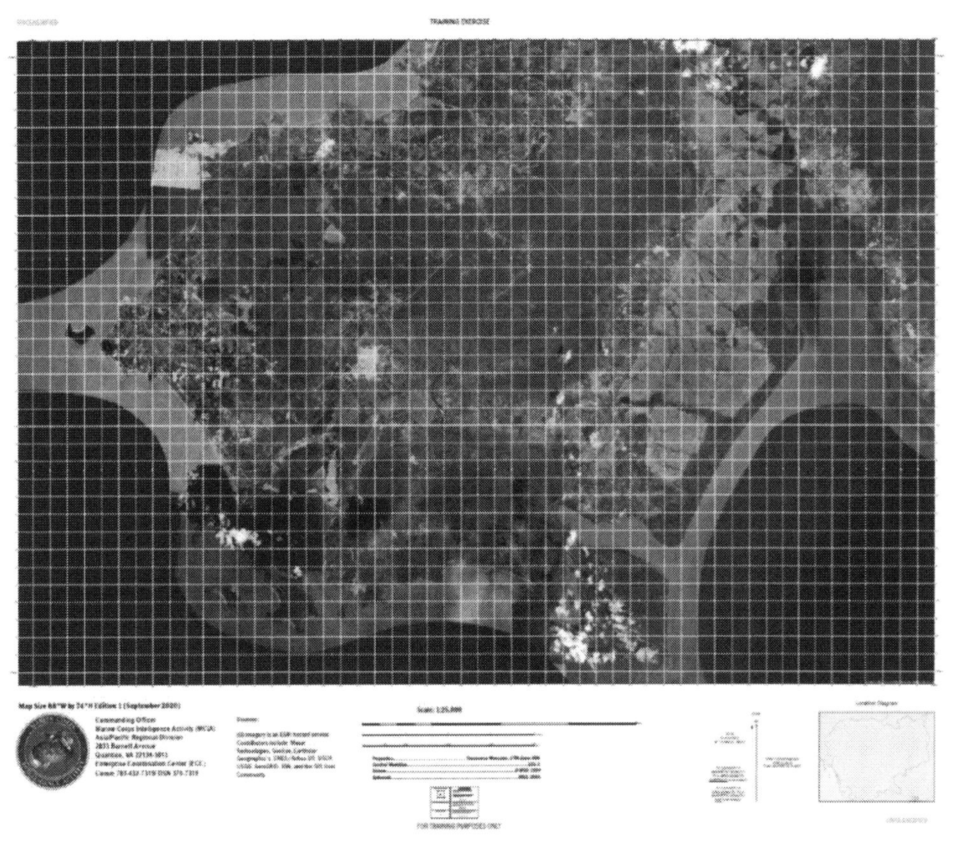

Photo credit: Nathan Rosenblatt, RAND.

side units and have icons in diamond patterns, in line with standard operational terms and symbology. Blue tokens represent BLUE units and have icons in blue rectangles, also in line with standard practice. Black icons with white backgrounds symbolize civilian groups and different types of information activities. Figure 16 depicts the general icon design from IWX 20.2, with red, black, and blue icons presented from left to right. Figure 17 depicts the large toy soldiers painted for IWX 23.3. Figure 18 shows the icons used for BLUE and RED units, formations, and capabilities in IWX 20.2, and Figure 19 shows the icons used for noncombatants and infrastructure in IWX 20.2. Finally, Figure 20 shows some of the actual tokens as used in the IWX 20.2 wargame.

More information on operational symbols and graphics can be found in Army Doctrine Publication 1-02, *Terms and Military Symbols*.

Each specific scenario will require a different mix of icons. However, the set acquired by MCIOC for the IWX 2020 wargame would be suitable for adaptation to many different scenarios involving Marine Corps forces and information activities. The RAND team acquired this set of icons from Litko Game Accessories (www.litko.net). Costs have already been paid for up-front artwork, so subsequent orders should have reduced costs. Anyone wishing to reorder a set of the IWX 2020 icons from Litko

FIGURE 16

General Icon Types Used in IWX 20.2

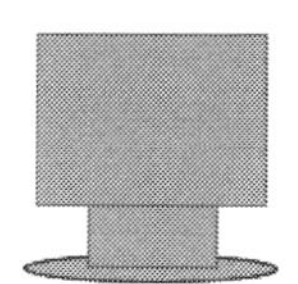

should reference order number 919268. Several other companies produce similar custom wargaming icons.

In 2023 for IWX 23.1, the MCIOC training cadre developed a new set of physical tokens. These were larger, in order to be less easily lost on a large plotter map on the floor, more stable, and less vulnerable to falling over. These were 4″ plastic soldiers that were fully painted in a bold blue or red color and affixed to wooden support bases. Each of the bases had a pair of holes drilled into it to support insertion of small dowel-based flags. These large figures were used to denote target audiences, with the flags added to denote successful influence.

Electronic icons: Each of these icons can be replicated in electronic format. Both SWIFT and VASSAL support the replication of icons in virtual form.

FIGURE 17

Figures and Markers Used During IWX 23.3

Photo credit: Thomas Schaeffer, MCIOC.

FIGURE 18

Icons Used for RED and BLUE Units, Formations, and Capabilities in IWX 20.2

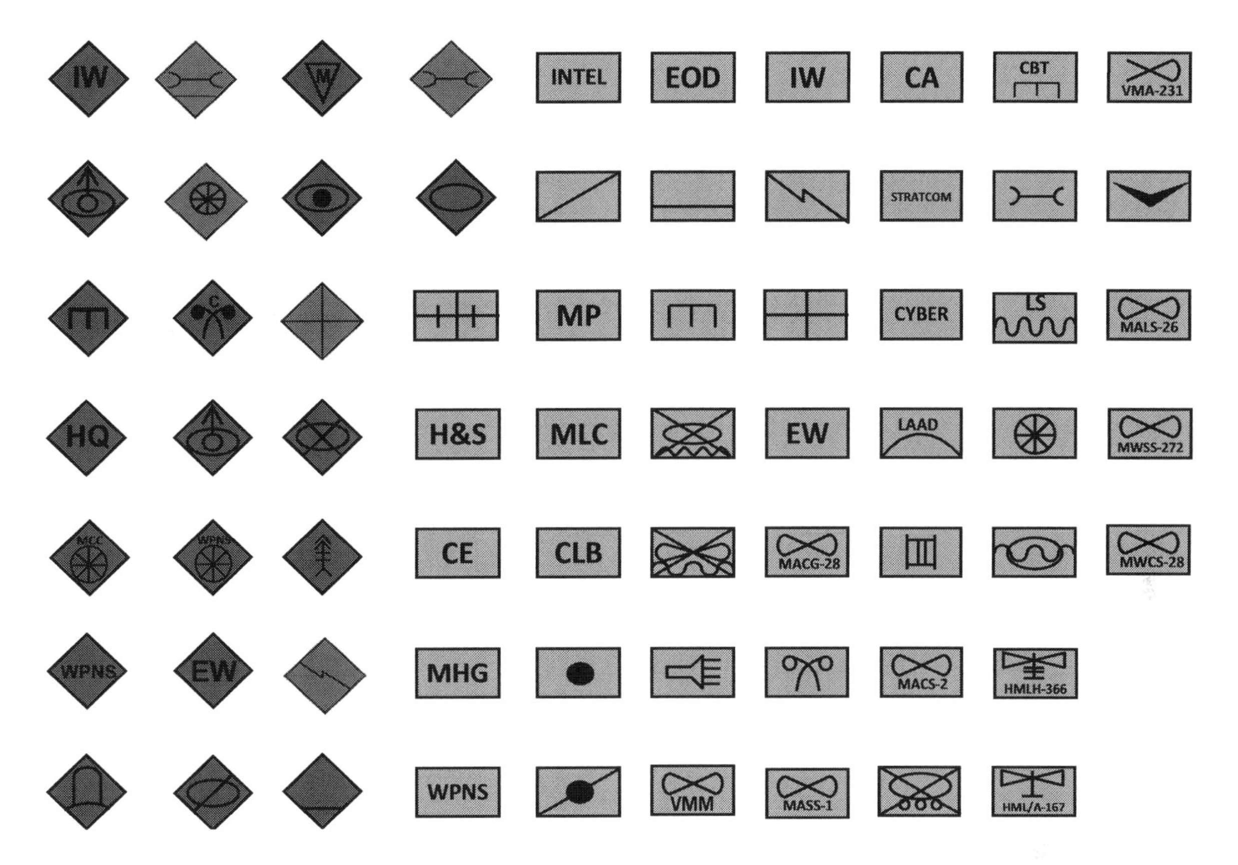

FIGURE 19

Icons Used for Noncombatants and Infrastructure in IWX 20.2

FIGURE 20
Physical Icons from IWX 20.2

Photo credit: Lance Corporal Kaleb Martin.

Additional materials: Playtesting and experience with IWX 20.2 revealed that additional markers and representations could be valuable. These might include translucent plastic radius templates (that could be used to show broadcast radii, jamming radii, drop zones for leaflets, etc.), as well as large plastic arrows to denote intended lines of advance (so tokens could show current positions of maneuver formations and relevant GREEN organizations or groups with large arrows showing their intended directions of progress during the turn, adding more dynamism to the battlespace). These arrows could correspond to the major colors in the game, so blue, red, and green. If using a laminated plotter map, these arrows can be created using dry-erase marker or grease pencil or could also be created using construction paper.

Also mentioned was the possible addition of "status rings"—plastic rings that can be added to the base of a token to denote some status of the unit (suppressed/pinned, out of communication, fleeing, resolved, etc.). These status rings might also be produced in a range of colors, and ring color might be used to denote the degree of influence one side or the other has over a unit (or civilian population group)—where no ring would denote normal/baseline degree of influence, a light red ring might denote some RED influence, and a bright red ring might denote more extensive RED influence (and the converse for BLUE), for example. Also, an observer suggested that flags or flag tokens could be made available and be used to denote BLUE or RED control over certain areas or key locations or facilities.

In addition, players in the wargame at IWX 20.2 found that having a pointer (a long stick, as shown in the back of Figure 13) was useful when briefing actions on the large game board/map.

3.8.1. Map Alternatives: Wall Map

Because of space constraints during IWX 21.1, EXCON was unable to display a large game board map horizontally on the floor or a large table. However, the space did accommodate a large plotter map hung vertically on one wall of the engagement room, which presenters were asked to stand in front of when presenting. This approach worked well and is a valid alternative to a horizontal map. Presenting players were able to refer to the map directly and were able to place and move sticky notes indicating the various capabilities and their activities. EXCON was able to note the position of other assets and to leave notes describing events from previous turns. Such notes were actually more effective than writing notes and attaching them to a floor map, as it is easier to read notes attached to a vertical surface at or above knee height than to read notes attached to the floor.

Seeing the large wall map adorned with information-laden sticky notes for the first time, one of the authors described it as "pin the tail on the donkey style," an apt characterization.

A vertical game board could be even more effective with additional materials and preparation. Any of the following, for example, could contribute further improvement: multiple colors of sticky notes, the capability to run a sheet of sticky notes through a printer and add printed text (rather than relying on handwriting of sometimes questionable legibility), or formal iconography. Actually having the map overlay a bulletin board or some other surface into which pins could be inserted would be a further improvement—sticky notes can and will occasionally fall off (especially if they've been unstuck and restuck repeatedly), whereas pins are much less vulnerable to accidental displacement. Furthermore, pins would allow the attachment of strings, and a single location on the map could have several strings running from it with sticky notes (or just notes or cards) pinned at the ends of the strings, clearly noting numerous capabilities or activities at that point (rather than a massive clump of partially overlapping sticky notes). In addition, pins and strings can provide good depictions and measurements of radii (for example, when a broadcast radius or jamming radius becomes operationally relevant).

3.9. Who Travels for IWX? IWX as Expeditionary for EXCON, for the Players, or for Both

The default approach to IWX presumes that the EXCON training cadre is on or near home ground, and that participants (and players) travel to that offsite location. What if that isn't the case? What if the IWX is conducted in an expeditionary fashion, and EXCON has traveled to the home station of the players?

IWX 20.2 and IWX 21.1 both included three-day periods of IWX Wargame play. However, they employed different scenarios, were conducted at different staff levels (tactical versus operational), and occurred in different locations and under different circumstances. These and other differences led to important observable variation in the flow of the two exercises and in the effectiveness of some exercise activities. Discussions with EXCON about observed variations and their impact also led to some speculation about future IWXs: If IWX 20.2 had one characteristic with one result, and IWX 21.1 had different characteristics with a second result, what might be the impact of a future IWX with yet a third characteristic in place? One of the topics of particular emphasis in these discussions had to do with the location of the IWX (and the wargame) and the relative location of the players and the supporting EXCON.

IWX 20.2 was conducted at Marine Corps Base Quantico, Virginia. All EXCON personnel were present live and in person, and all exercise participants were also present in person and had traveled from their regular duty station or office to attend. IWX 21.1 was conducted at Camp Smith, Hawaii, in spaces belonging to U.S. Marine Corps Forces, Pacific. A cadre of EXCON personnel traveled from MCIOC (following a mobile training team model), and various additional MCIOC personnel were available via Secret Internet Protocol Router Network (SIPRNet) video teleconferences (S-VTC) for reachback support. Exercise participants were in the same facility as their duty station, with the attendant pressures and temptations of their regular workstations and duties. The authors have participated in or observed numerous other exercises and training events conducted under a wide range of circumstances and in a variety of styles, which further informed speculation about how different approaches might affect an IWX or an IWX Wargame.

3.9.1. When EXCON Travels

These experiences led to several observations. Among them, it is better to have as much of EXCON present in the exercise location as possible. Reachback was adequate for some exercise requirements (such as answering requests for information or delivering a block of instruction), but advising and mentoring participants (especially during the wargame) requires a level of connection and rapport that is much easier to build while on site. An exceptional mentor or someone with whom many of the participants already have an existing relationship might be able to manage that role remotely. If an IWX Wargame is conducted on a mobile training team model, it might be best to recruit mentors and advisors from among personnel local to the exercise locale and have them on site for the exercise.

A mobile training team approach is often called for when a stakeholder simply cannot afford to allocate two full weeks of training and travel time for key personnel. Thus, an expeditionary IWX is also almost always a shorter IWX. See the discussion in Section 4.2.

3.9.2. When Players Are at Home

The flip side to EXCON traveling is that unless the IWX is off-site for everyone, the participants/players will be at home. Experience has revealed that there are benefits to a captive audience. If participants have traveled at least some distance and are away from their own duty spaces, then the exercise and the wargame have their full attention. This is especially true for exercises that take place in secure spaces where access to personal electronics is constrained. When participants are in or near their own desks and duties, the temptation to split time between the exercise and those duties is strong. This might be beneficial to the accomplishment of those duties, but it is not beneficial to achieving all the learning objectives of the exercise. **Where possible, participants should be removed from the temptation of their day job**, either traveling as part of temporary duty to another location, or moving to a nearby location that is still "off-site"—that is, not in or within easy walking distance of their daily workspace.

If conditions require that an IWX Wargame take place within participants' regular daily workspace, steps should be taken to capture their attention for the game. This might be accomplished in several ways. One approach would be a clear communication at the beginning of IWX by the formation commander/office manager that participants are excused from their regular duties for the duration of IWX and are expected to devote their full attention to the exercise and wargame. Another approach might involve participants and EXCON making a clear public commitment to each other to devote their full attention to the wargame during the wargame period. This latter approach could be bolstered by a symbolic gesture, either a group pledge of attention and avoidance of distraction read together by all participants at the start of the wargame or something like "stacking phones," whereby all participants signal their commitment to be present and avoid distraction by making a pile of mobile phones (the principal distractor in many work and social situations) and giving their full attention to the group activity.[9] Similarly, players might be dissuaded from routine duties through a system whereby players caught at their own workstations receive symbolic demerits or pay a nominal fine, perhaps $1 each time "caught," with proceeds collected supporting a nicer lunch on the last day of the exercise. Under that model, if the pool of fines collected is small (indicating good levels of attention capture for the game), EXCON representatives or formation leaders could contribute to funding the nicer lunch as a reward.

[9] In secure workspaces where mobile devices are already restricted, some other symbolic gesture would be required.

4. Optional Rules for Scaling the IWX Wargame

The IWX Wargame is designed for 12–24 players playing five turns, and for this endeavor to consume a significant fraction of a two-week exercise period. However, IWX (and the wargame) has taken several different forms and might take further different forms in the future. In addition to the flagship two-week IWX in the fall, MCIOC often conducts a one-week expeditionary IWX in response to an individual command's exercise request. The growing popularity of IWX increases demand for available participant slots during the flagship event. What might be done to accommodate more players? What might be done if the time available for the entire exercise is one week rather than two? This section addresses considerations related to scale, either increases in the number of participants or decreases in the time allocated for the game.

4.1. Options for Accommodating More Players

Like most exercises, IWX announces that it has a certain number of slots available for participants. Like many events, there can be a waiting list, and the MCIOC training cadre can accept a few more participants than there are actually slots for with the expectation that at least a few of the individuals who have announced their intention to attend will fail to do so, either due to competing mission critical tasks, illness, or other misfortune. Still, as demand for slots in IWX increases, what, other than holding an IWX more frequently, can MCIOC (or a future different IWX Wargame user) do to increase the number of players that can participate in the wargame?

4.1.1. Considerations When Accommodating More Players

There are two main approaches to accommodating additional players (described in Sections 4.1.2 and 4.1.3): (1) either absorbing additional personnel within the normal confines of the game by somehow creating additional player roles or subteams or (2) running multiple instances of the game. The biggest cost in terms of either approach is in units of *time*.

The biggest and least flexible time consumer within the game is adjudication, with a single action costing approximately 30 minutes, and so a paired cycle of one RED and one BLUE action costing approximately one hour. If additional players are absorbed in a way that still leaves only three actions per turn per side, then the impact on the overall game is probably minimal and manageable. If, however, added subteams receive *additional* actions rather than just an allocation from the actions already available, it will increase the overall time required to complete a turn cycle. This might be managed with reductions in time allocated to other steps (such

FIGURE 21

A Large Team of Players Work Together During IWX 23.3

Photo credit: Jonathan Welch, RAND.

as planning; perhaps more subteams can plan more actions more quickly?) or by keeping the training audience for longer than a typical 8-hour day.

By contrast, adding a second instance of the game preserves the timing and flow of each game, but the total time required to adjudicate actions (a total of 12 actions per day, three actions each from two RED teams and two BLUE teams) would really balloon. While this time can be simultaneous (a different adjudication process for each game), the extra time still must be paid by EXCON. This could either be accommodated by an entire additional set of EXCON positions and roles, or the same EXCON group running both games, jumping back and forth between them and the judging panel (for example) running Step 4 for one game in the morning, and then turning around and doing adjudication for the other game in the afternoon. Either of these alternatives is taxing, requiring another full EXCON cadre or taking the typical cadre and demanding an extremely heavy level of effort.

When considering ways to accommodate additional training audience members, game hosts should certainly consider the time required to fully complete each turn and the additional burden in EXCON person hours of effort required to support the expanded game or games.

4.1.2. Expanding by Subdividing: Finding Things for Additional Players to Do

More players can be accommodated by finding additional roles. A certain amount of extra capacity can be had through simply adding extra players to teams and being slightly over optimal size. The first edition of this ruleset sought a cap of 10 participants per team; the IWX Wargames that have been conducted since publication of the first edition regularly allowed 12 players per team without disruption, so that is now the target cap. This might be stretched to as high as 15, provided the physical spaces for the teams and for adjudication will accommodate that many players on each team. 15 is an important threshold, as with five turns of three actions each, all 15 players would get to present exactly one action each; any players beyond 15 and someone misses that opportunity.

Another way to accommodate additional players is to formally subdivide the teams. This would require additional planning space (either an additional planning room for each subdivision, or a team planning area that will accommodate all team members and allow for two separate activities to be taking place without causing excessive disruption). With such space available, several types of subgrouping or subdivision are possible:

- One or more subteams, such as a deception working group, a signature management working group, or a subteam focused on civil affairs and humanitarian efforts
- A special operations forces (SOF) layer, with separate planning to support the actions of an attached SOF formation for one or both sides (this would require some additional scenario development work to make sure there is sufficient planning information for these SOF elements' intended actions)
- A division of the team by echelon, with one subteam playing at the tactical level and the other playing at a higher, operational echelon (see Sections 5.1 and 5.2 for considerations related to playing the game at the operational level).

Again, the principal consideration when subdividing a team to accommodate more players has to do with how many actions each subdivision will be allowed to take and the total number of actions allowed. If, for example, a subteam is allocated one action per turn, and the core team retains the other two, then the total actions per side per turn remains at three, the game's timeline remains manageable, and Step 4 can still be resolved in a single morning or afternoon of play. Alternatively, if a subteam's actions are largely separate from and largely unobserved by the core elements of the two sides (as might be the case with SOF), then a separate mini-adjudication could be run on the side by a subset of EXCON perhaps while the rest of the players are doing Step 3, and then any effects or consequences from those actions that rise to a detectable level can be mentioned when appropriate during Steps 4 or 5. This is not ideal, because it changes the matrix debate structure and the communal learning afforded by having everyone present for adjudication, but it would accommodate more players.

Finally, additional participants do not necessarily have to equate to additional players. Relatively senior participants might be recruited into nonplaying roles in the game, such as senior mentor or EXCON roles, perhaps serving as part of the panel of judges.

4.1.3. Expanding by Separating: Multiple Simultaneous Games

The options presented in Section 4.1.2 preserve the game as a single instance. Multiple simultaneous games would allow for the accommodation of considerably more players, but at a notable cost in EXCON time.

Under this approach, the exercise would add whole additional BLUE and RED planning teams. These additional planning teams would have their own "parallel universe" in which their actions (and the storyline) would unfold. Both separate games would start from an identical starting point but would diverge based on the different actions taken by the teams in the separate games and the different outcomes of those actions. As noted, running two games would require either a full additional set of EXCON training cadre (and spaces, and game materials) or a heroic effort from a single EXCON crew, which would demand additional time and create challenges related to keeping track of what things had taken place in which scenario universes.

Another option for doubling the number of teams would be to have two separate planning cells for each side in the game, describing them as adjacent formations at the same echelon across a boundary. This approach preserves a single-scenario universe (i.e., everything that happens all happens in the same universe) but creates other concerns and considerations. Paramount is the total number of actions. If each planning team gets three actions, adjudication would basically take all day. This would be untenable unless players are asked to plan their actions overnight or turns are stretched to multiple days, with a half-day of planning preceding each full day of adjudication. Alternatively, if each planning team is allowed two actions rather than the typical three, the total number of actions per side is four, a large but not unmanageable number if time is trimmed slightly elsewhere in the process.

Other considerations related to running multiple planning teams per side include the physical space to accommodate additional planning groups and a requirement for some kind of effort to maintain verisimilitude regarding deconfliction and coordination between the two teams on a side. Two separate formations' planning staffs would be located in separate headquarters or command posts, not in the same room or just across the hall. Deconfliction and coordination would have to be done through remote means. To preserve at least some realism in this regard, planners on the same side but different teams should be forbidden from speaking face to face, instead required to do any coordination via email, chat, phone, or VTC of some kind.

4.2. Shorter Versions of the IWX Wargame

It can be difficult to carve out two weeks to conduct a full IWX, especially for higher-level staffs. However, this need not eliminate the IWX from consideration as an option for accomplishing the objectives identified in Section 1.2. There are participant considerations, prework options, and ways to abbreviate the game itself that can make it accessible when its full form is not an option and a full week is not available for the game.

There are several player considerations that could affect the smooth running of the game, and these might, in turn, influence the selection and assignment of players as addressed in Section 3.1. First, the experience base of the participants is a key factor in how efficiently a shortened game runs. Veteran players (that is, players who have played the IWX Wargame before) need not spend as much time learning how to play the game and can instead focus on understanding the scenario and moving into planning. No two IWXs are alike, though there are pros and cons to playing the exact same scenario repeatedly. The takeaway for EXCON is to consider IWX experience when building the RED and BLUE sides. If there are multiple IWX veterans, they ought to be dispersed between the teams.

Second, repeated IWXs have shown that planning experience, in the general sense, pays dividends for players and their teams in terms of efficiency. This is a distinct skill set from tactical information activity planning, and it does not have to be experience with the Marine Corps Planning Process; experience with any service or joint planning process will do.

Finally, team composition matters. Typically, but not always, teams are composed of a variety of players, which may include individuals from not only multiple units but multiple military services and multiple nations. Obviously, a trained and experienced information planning team that has worked together as a team ought to move through the exercise effectively and efficiently, regardless of their experience with the IWX Wargame format. Again, experience should be considered when building teams. Decisions can be made in favor of efficiency, but this may come at the cost of exposure to other ways of doing things.

MCIOC or another IWX host should be aware of the above participant and team considerations when designing a shortened IWX Wargame, even if they have a limited ability to change them. Still, there are some things that the host can do, in coordination with participants and their parent organizations or commands, to gain a bit of efficiency in the game before condensing it. One option is to load preparations for the wargame prior to the game beginning. Read-ahead scenario materials and planning sessions before the IWX can reduce the time needed for planning and familiarization during the exercise itself. Unfortunately, many exercise participants are not naturally inclined to read ahead or preplan unless properly incentivized. Two possible incentives are (1) commanders making preparations a priority (and setting aside time in the workday for such preparations) and (2) stimulating players' competitive spirits by pointing out that members of the other team are already planning and that they should do so as well to avoid being disadvantaged when the game starts. The bottom line is that individual, or better yet group, investment in gaining familiarity with the scenario prior to physically attending IWX can decrease time spent learning it in residence.

Similarly, a basic familiarity with how the IWX mechanics function can save time for focusing on IWX play. This is one of the key reasons for the *Player's Guide* that accompanies this rulebook. Building familiarity with the scenario and ruleset prior to IWX attendance can be accomplished virtually, or in person, but is best done as a group. The added benefit of this sort of prework is that it presents an opportunity for players to get to know each other a bit and do some preliminary team-building. Finally, taking it one step further, the steepest planning cost in the IWX is Step 2 in turn 1. It takes some time for a new team to work their way through building out their first set of actions, and the game flows more smoothly after that. Executing this step prior to IWX attendance could save substantial time in residence.

NOTE: These hybrid options of virtual preparation preceding in-person play are akin to how many professional military education courses are now delivered and should be familiar to many participants. However, overcoming connectivity and classification challenges requires planning and effort. Also, although it is technically feasible to conduct an IWX entirely virtually, we strongly recommend avoiding this option. Planning may be conducted virtually, but the in-person experience in the Engagement Room, in Step 4, is simply irreplaceable.

Finally, if an abbreviated game is a must, there are a couple of ways to condense it, albeit with a likely degradation in training value.

Reduced Turns. The simplest way to shorten the IWX is to cut the number of turns. As noted in Section 1.3.2. a typical turn takes roughly 8 hours. This means that cutting a turn saves a day, but it comes at the cost of a valuable repetition. Anything less than three turns is inadvisable, as it does not allow for the scenario to build sufficiently or for players to get enough rounds in to gain proficiency. However, with three turns, it is still possible to run the IWX Wargame in the course of a five-day work week, with requisite preparations on the front end and a comprehensive AAR at the conclusion. It may be possible to slim it even further with prework, as suggested above.

Reduced Actions. Another way to abbreviate the IWX Wargame is to reduce the number of actions. This affects Steps 2 through 4, but Step 4 most significantly. Marginal time savings might be gained by planning fewer actions or pursuing their approval in Steps 2 and 3, respectively. However, at 30 minutes per action (Section 2.4.5), cutting two actions (one from each team) could save about an hour. Given the broader flow of the IWX, it would be difficult to reduce the number of days by reducing the number of actions. However, reducing the number of actions might make the IWX more accessible to staff members unable to fully disengage from their day jobs.

Note: A possible concern when the number of turns or the number of actions is reduced is the increase in the vulnerability to randomness in the dice. The law of large numbers demands that over a large enough sample, rolls of dice will average out and cover a Gaussian (normal) distribution. However, each individual roll is fully vulnerable to the whims of chance. As the total number of dice rolls in the game declines, the risk of many of those rolls being in the tails of the distribution (abnormally high or abnormally low) increases. In the full game, with 2 rolls to adjudicate each action, 6 actions per turn, and 5 turns, there are a total of 60 rolls in the game (30 for each team). If instead each team only gets 2 actions per turn over only 3 turns, the total number of rolls falls to 24 (only 12 total per team). If the dice behave, this could be fine, but a reduced game is simply more vulnerable to chance.

5. Optional Rules for Conducting the IWX Wargame at the Operational Level of War or for a Competition Scenario

The IWX Wargame rules are scenario-agnostic and were designed to apply to a wide range of possible operations and contexts. However, the original scenario, used to fine-tune the wargame rules, involved a conflict in which BLUE (the Marine force) was tasked with the seizure of a port and airfield against the opposition of a similar formation of RED (opposition) forces. Operations, both the scenario-driven maneuver effort and the information actions planned by the players, took place at the tactical level and involved battalion-sized formations and appropriate supporting elements.

In 2021, MCIOC conducted an IWX (IWX 21.1) for U.S. Marine Corps Forces Pacific (MARFORPAC). Tailoring the exercise and wargame to the MARFORPAC training audience involved several differences from the original baseline IWX Wargame. First, the training audience was an operational-level headquarters rather than a tactical unit staff, so MCIOC chose to have the exercise teams plan and play at the operational level.[10] Second, in consultation with MARFORPAC, MCIOC determined that a competition scenario, rather than a conflict scenario, would provide the most useful exercise experience.[11] The authors supported MCIOC in applying the IWX Wargame to a competition scenario played at the operational level.

This chapter provides supplemental rules and considerations relevant to conducting an IWX Wargame at the operational level, using a competition scenario, or both. It also presents some observations and lessons learned from the playtesting and execution of the IWX Wargame for MARFORPAC.

5.1. Considerations and Challenges Related to Conducting an IWX Wargame at the Operational Level

What is different about planning and staffing information activities at the operational level, and how does that need to be reflected in the IWX Wargame? Discussions with and among various MCIOC personnel exposed at least two important differences between operational-level and tactical-level planning. First, the activities of an operational-level headquarters are focused more on "up-and-out" coordination and deconfliction with a host of relevant stakeholders, including adjacent commands, higher headquarters, interagency elements, and international partners, whereas tactical staffs have a much narrower range of headquarters to coordinate with and focus much more on directly coordinating information activities and the details of those activities with those actually executing them. Second, operational-level planning involves identifying and requesting

[10] According to joint doctrine (Joint Publication 3-0, 2022), "The framework of strategic, operational, and tactical levels of warfare helps commanders visualize the relationships and actions required to link strategic objectives to campaigns and major operations and link their objectives to tactical operations." The strategic level deals with the pursuit of national-level interests through the application of the national instruments of power, often framed in the "DIME" construct: diplomatic, information, military, and economic. It is important to note that U.S. national interests often overlap with those of other nations, so a key part of strategic planning involves multilateral coordination and cooperation. The operational level "links the tactical employment of forces to strategic objectives." This is the level at which planners apply "operational art" to devise ways to achieve high-level objectives with available resources. This is where component commands do most of their work. The tactical level is where forces focus on "planning and executing battles, engagements, and activities." The tactical level is the bread and butter of Marine Corps units.

[11] On the competition continuum, cooperation characterizes activities at the most benign end of the continuum, and conflict is the most aggressive end and involves open armed hostilities. Between the two—and occupying much of the spectrum—is competition, which can include a wide range of different activities at different levels of aggressiveness but is designed to stay below the threshold of open warfare. This competitive space is sometimes referred to as the "gray zone." For a further discussion, see Lyle J. Morris, Michael J. Mazarr, Jeffrey W. Hornung, Stephanie Pezard, Anika Binnendijk, and Marta Kepe, *Gaining Competitive Advantage in the Gray Zone: Response Options for Coercive Aggression Below the Threshold of Major War*, RAND Corporation, RR-2942-OSD, 2019; also see Marine Corps Doctrinal Publication 1-4, *Competing*, U.S. Marine Corps, December 14, 2020.

authorities and permissions and providing guidance for subordinate elements to execute within, whereas tactical planning produces specific and detailed planned actions. Both of these differences require consideration in the design and execution of the IWX Wargame.

The up-and-out focus and coordinating activities of an operational-level staff require greater attention within the game to the complexity of relevant levels of approval and interagency stakeholders. Whereas a tactical-level staff can proceed with the employment of organic capabilities based on the approval of the plans by the S3 or the commander, an operational-level staff needs to coordinate with, consult with, identify authorities through, or gain permission from a wider range of stakeholders. In fact, identifying the appropriate commands and organizations with which to coordinate and understanding the processes for such coordination (including timelines, etc.) are important learning objectives that are distinct to an operational-level IWX.

This raises at least two issues for the IWX Wargame for an operational-level staff. First, EXCON and the IWX Wargame structure need to include ways to represent or include a much wider range of stakeholders than what is included by default. Second, questions regarding coordination are both more likely to come up and less likely to be quickly and easily resolved than during tactical-level play. These coordination questions and discussions are important to successfully meeting learning objectives at this level and are time well spent, so the wargame structure needs to allow more time for these discussions to unfold.

Because operational-level planning produces guidance to structure the actions of subordinate units, operational-level actions (one of the core building blocks of the IWX Wargame) are necessarily more abstract than tactical-level actions. The IWX Wargame action scoresheets (see Section 2.4.3.8) might need to be changed or applied slightly differently to effectively evaluate and generate action target numbers for actions based on operational-level planning.

5.2. Considerations and Challenges Related to Conducting a Competition-Focused IWX Wargame

In the same way that operational-level information planning differs from tactical-level information planning, scenarios that include operations in competition differ from those during conflict. The literature on competition is voluminous,[12] but discussions with MCIOC personnel and observations from playtests and the actual execution of the IWX 21.1 wargame revealed only a few differences that are consequential for the IWX Wargame. First, because competition seeks to incrementally gain influence while avoiding escalation, play and actions during a competition scenario bias toward risk-aversion. Second, and related, competition actions tend to seek plausible, relatively mild effects and so are "easier" to execute and more likely to succeed. Third, actions during competition can be subtle, nuanced, or complicated, which can also make it harder to measure their effectiveness and can increase the plausible time gap between the start of the action and its effect (if any) becoming observable.

All these differences suggest adjustments to the IWX Wargame, either in the rules, in scenario design and exercise planning, or in execution by EXCON. First, although "safe" and risk-averse actions are realistic, they make for a less exciting wargame and are less likely to stimulate the full complexities of the processes of information planning. A tendency toward risk-averse play could be left to run its course or could be disrupted through scenario guidance, requirements that necessitate risk, or the addition of injects that provide needed stimuli. Second, the modest effects sought in competition tend to make competition wargame actions seem easier than tactical actions; the scoresheet and

[12] For one review, see Michael J. Mazarr, Jonathan S. Blake, Abigail Casey, Tim McDonald, Stephanie Pezard, and Michael Spirtas, *Understanding the Emerging Era of International Competition: Theoretical and Historical Perspectives*, RAND Corporation, RR-2726-AF, 2018. For another, see Christopher Paul, Michael Schwille, Michael Vasseur, Elizabeth M. Bartels, and Ryan Bauer, *The Role of Information in U.S. Concepts for Strategic Competition*, RAND Corporation, RR-A1256-1, 2022. For specifics on the Marine Corps view of competition, see Marine Corps Doctrinal Publication 1-4, 2020.

system for determining an action's target number (see Section 2.4.3.8.1) might need to be adjusted to prevent too many actions from succeeding too spectacularly. Finally, understanding and exploring the subtleties of competition actions can require more time, both to plan and to explain and discuss sufficiently for adjudication (and learning); IWX Wargame timelines and workflow can be adjusted to accommodate.

5.3. The Importance of Objectives in Both Competition and Operational Wargames

One of the observations from playtesting for IWX 21.1 and the exercise itself as a playtest was the increased importance of objectives. In a tactical and conflict-focused scenario, there are clear maneuver objectives for each side for each turn and for the overall operation provided by the scenario through the medium of the commander's guidance or the instructions of the S3 role-player. An example might be "B Company will advance down Route Epsilon to the airfield and secure it." This in turn leads to information activities with clear connections through supporting objectives; for example, several information capabilities might be used in concert to reduce civilian traffic or other obstructions on Route Epsilon, or a deception effort might have the objective of drawing defenders away from the airfield or causing them to orient their defense in a different direction. Even though maneuver objectives can change rapidly in a dynamic tactical environment, they remain clear, and it is relatively easy for teams of participants to align information activities to support of those objectives. When information activities stray, it is easy for EXCON mentors or role-players to steer the team back on course: "Wait a minute . . . Bravo Company is about to advance toward the airfield on Route Epsilon. Why are you still talking about jamming comms at the port?"

This clarity and focus and ease of nesting information activities to support maneuver actions and objectives proved to be less automatic in an operational-level competition-focused scenario. Vague (or just broadly scoped) initial guidance and operational objectives can lead to vague or imprecise information actions that do not clearly and neatly nest within those broader objectives. "Why are you doing this?" and "What do you hope to gain through this?" became more common questions from EXCON to the players. These questions often had answers, but sometimes these answers were not fully satisfactory, and the nesting and connection to higher-level objectives was not always obvious.

This phenomenon was observed in playtests and in a wargame instance that was both operational-level and competition-focused. On reflection, we believe the competition scenario was the prime driver. Competition covers a broad range of possible goals; operations during competition might seek to make miniscule gains against, or to set conditions favorable to, many of those goals, or perhaps to exploit an adversary's misstep or a contextual opportunity. The same risk-aversion noted above as being common in competition also promotes less clear and precise initial guidance and operational goals. Future IWX Wargame competition scenarios might include several characteristics that could mitigate against this challenge and promote stronger information activities with clearer connection to broader operational objectives. Scenario design can help with this; see Section 3.4.1.1.

Finally, when exercising a component command currently engaged in "real-world" competition, it is important to be explicit about what is "real-world" versus what is specific to the exercise scenario. For example, component planners are familiar with standing operational plans (OPLANs), execution orders (EXORDs), and other guidance from the geographic combatant command. These provide a wealth of objectives (perhaps even an overwhelming number). To effectively exercise a scenario, clear boundaries need to be drawn between what is occurring in the real world and what guidance and events are specific to the fictional future scenario.

5.4. Involving Multiple Staff Layers as Players

Competition or operational-level games might create a desire to represent multiple echelons formally as players. However, Section 1 describes the team of players for each side as representing a single information working group or OPT, with the assumption that staff sections at other echelons or elements will be represented, to the extent necessary, by EXCON. The optional rules in this section allow for the possibility of having players, rather than EXCON, portraying additional command echelons. These rules can accommodate players representing up to three staff layers, including a tactical-level staff echelon, such as a Marine Expeditionary Force (MEF) staff; an operational-level staff echelon, such as a geographic component command like MARFORPAC; and a theater-level echelon, such as staff from a geographic combatant command or combined joint task force.

5.4.1. Theater-Level Participants

Including participants from higher headquarters or the interagency can meaningfully enrich the IWX Wargame experience. These individuals can help to ground the game in the realm of the possible while providing a more nuanced understanding of how the subordinate command should interface with other layers and organizations. Similarly, those additional participants are likely to walk away with a better or renewed perspective of the challenges faced by the primary training audience.

5.4.1.1. Full-Time Versus Part-Time Participants

A primary consideration in determining whether and how to include additional staff layers within the IWX Wargame is the intended level of participation of the additional participants. Tactical or operational staffs should likely be represented by full-time participants in the IWX Wargame, but the need for theater or similarly high-level staff echelons to be represented might be episodic: The higher echelon(s) might have an important role in certain scenarios, but once they have provided needed permissions, identified relevant authorities, issued planning guidance, or simulated high-level coordination processes, much of the game can proceed without them. If such a staff layer is represented by participants, they can be full-time IWX participants (see the next section) or part-time participants, either on call or available and engaged only during specific preplanned windows (see Section 5.4.1.1.2).

5.4.1.1.1. Full-Time Higher-Echelon Staff

If a higher-level staff is represented by participants who are present during the entire IWX Wargame, they can be part of both EXCON and the training audience. Functionally, for the wargame, they are part of EXCON; however, the interactions they have with the subordinate staffs and the likely gain in understanding of various processes at other echelons or in other organizations should be very valuable to them as learners/trainees. Participants in this role would not have their own separate workspace but would instead "reside" in the EXCON workspace, visiting the workspace for their assigned team as appropriate. During Steps 1 and 2, higher-echelon representatives should be available for RFIs from subordinate staffs or to provide guidance related to process at their echelon. During Step 3, higher-echelon representatives can take the place of the S3 role-player and receive their team's approval brief (Section 5.4.1.2) and approve or disapprove actions (Section 2.3.4). Higher-echelon staff filling this role should report back to EXCON on the actions they approved (Section 2.3.4.4). Also, during Step 3 and subsequent to action approval, higher-echelon representatives can support the EXCON narration and scenario team in the preparation of narration for possible action outcomes (Sections 2.4.3.8.2.2 and 2.4.3.9.2). During Step 4, higher-echelon representatives can be present and able to respond to EXCON or team questions about processes appropriate to their echelon but would not have a role in the adjudication process.

5.4.1.1.2. Part-Time Higher-Echelon Staff

If representatives of higher-echelon staff (potentially including various interagency representatives) are only available for part of the wargame, they can still make valuable contributions. Depending on the extent of their availability, they could be scheduled to participate in various parts of the wargame process either at pre-scheduled intervals or on an "on-call" basis. The priorities for higher-echelon participation would be

1. sitting for a pre-approval/confirmation brief (see Section 5.4.1.2) prior to the first full turn of the IWX Wargame
2. joining Step 4 of each turn to answer questions about processes at their echelon (and observe the actions and related discussion)

3. being on call for RFIs or process questions from the teams during Steps 1 and 2
4. replacing or supplementing the S3 role-player and receiving the action approval brief during Step 3 (see Section 2.3.4).

NOTE: When including part-time participants, there is always a certain amount of catching up for them when they return to the game. With that in mind, it is best to have them present and available from the start of a step or instructional or interactive segment so that they can be caught up once, right at the beginning. When organizing part-time participation, remember that while having them present for a full step or segment is best, it is usually better to have intermittent participants leave early than to have them arrive late. Plan accordingly.

5.4.1.2. Theater-Level Panel Pre-Approval/ Confirmation Brief

Particularly in a competition scenario in which rapid crisis or wartime authorities are absent and information activities need to be planned and approved well in advance, one option for involving representatives from higher-level staffs (or interagency representatives) is to form a panel of such representatives and have them receive a confirmation brief of the team's entire plan before the start of the wargame. This "turn 0" event could be the culmination of the transition between the IWX planning period and beginning the formal IWX Wargame. See Sections 2.3.9 and 2.3.9.1 on using a confirmation brief and a confirmation panel.

5.4.2. Including Both Tactical and Operational Staff Layers

Including theater-level representation might be attractive, but it might also be useful to exercise the interface between tactical and operational staffs. If this optional rule is included, both tactical- and operational-level staff participants should be full-time participants.

One option for including a tactical staff (perhaps a MEF information group [MIG] team) and an operational-level staff (such as MARFORPAC) would be to include a full team representing each staff and assign each staff its own planning room. Under this construct, the staffs would visit each other's planning room only temporarily, to represent communication between the staffs.

The preferred option, however, is to have the two staffs playing as part of a single group and sharing the same team space, but have each player clearly assigned to one of the two echelons (tactical and operational). Under this construct, the operational staff would provide guidance, note processes and staff mechanics, and take the lead in any engagement with higher echelons (whether represented by players or role-players), and the tactical staff would focus on preparing the tactical actions and coordination of the information capabilities.

Inclusion of multiple full-time-participant staffs as teams has implications for the spaces required for the wargame. If each staff is assigned its own workroom, then additional rooms are required. If multiple full-sized staff teams are used, then the space used for Step 4 will need to be larger. Section 1.5 specifies 6–12 personnel per team. This is a good size for a working group: It allows everyone to participate and is easily accommodated in most conference rooms or workrooms. The division of the participants into two staffs for a team, whether in separate rooms or sharing a space, might increase the number of participants beyond 12, but we caution against fully doubling it. 15 is a reasonable target combined staff size limit—18 people might be a maximum manageable size, and 20 would almost certainly be too many.

5.4.2.1. Action Approval Between Staff Layers

If multiple staff layers are represented, there is the option of having one staff layer preside over the Step 3 action approvals for the subordinate staff layer. This was already noted in Section 5.4.1.2 as an option when including a theater-level echelon. This could also be used with an operational-level staff receiving and approving the action briefs from a tactical staff. Several considerations:

- A member of EXCON should still be present and chair the meeting to make sure that proposed actions satisfy game requirements (Section 2.3.4.1) and to be able to inform the rest of EXCON about what actions are planned to support preparation of initial scoresheets and outcomes narration (Section 2.3.4.5).

- If the operational-level staff and tactical staff are playing as separate teams in separate workspaces, it is recommended that the operational-level team be the approvers in Step 3; providing guidance and ascertaining whether that guidance has been met becomes a major part of that staff's role in the game.
- If the operational-level staff and tactical staff are playing as a single team sharing the same workspace but having slightly different roles, it is recommended that the operational-level team not be the approvers in Step 3. In this case, the operational-level team will already be fully aware of the details of the plans prepared by the two staffs working in concert, and the Step 3 brief would be unlikely to communicate new information in either direction. Instead, an EXCON role-player or EXCON-affiliated higher-echelon representative should conduct the Step 3 approval.
- If an EXCON role-player conducts the Step 3 approvals, some consideration should be given as to whom that individual should represent. In the default game, aimed at the tactical level, that role-player represents the S3 from that side's command. That may or may not be the right approval authority for a combined operational/tactical plan, depending on what capabilities it entails. One possibility is to leave the EXCON approver's role-played position initially undefined. Then discuss, in Step 3, the processes that would be necessary and the level of approval authority that would be required to get the team's plan approved. At that point, EXCON could announce that its role-player represents all (or some subset) of those processes and approval authorities. That role-player can also withdraw to their "higher headquarters" (the broader EXCON team) to consult with them on how to deal with unique issues and return to the team with "the feedback they got from higher."

5.4.2.2. Presenting Actions Prepared by Multiple Staff Layers

Because the core of the IWX Wargame's adjudication mechanism revolves around team actions, these must remain at the core of the adjudication process in Step 4, regardless of what and how many echelons are participating. If many staff layers are represented in the wargame, game actions will be presented in Step 4 by the lowest staff echelon represented.

If the game includes players representing both a tactical and an operational level staff, there are two options for adjustments to the Step 4 process described in Section 2.4.3.

The first option follows the default rules almost completely, and the tactical echelon players make the presentations and engage with the opposed teams. The operational echelon players are present in the engagement room and are available to answer echelon-appropriate questions from EXCON, but they otherwise have no role beyond observation; their main role is in providing guidance during Step 2 and in approving the actions in Step 3. This approach has the disadvantage of giving some of the participants no significant role in the most exciting part of the wargame, and so is not the recommended approach.

A second option is to have two presenters—one from the operational staff and one from the tactical staff—give a *hybrid presentation* for each action. The presenter from the higher echelon begins the presentation and is responsible for topics related to staff mechanics and guidance given, then the subordinate echelon presenter gives execution details. Specifically, the higher-echelon presenter would brief task, purpose, and end state, then the subordinate-echelon presenter would brief methods, execution details, and MOEs, as well as presenting three reasons the action would succeed. (Either presenter could offer reasons the action would succeed, but assigning that task to the subordinate-echelon briefer limits the number of handoffs or transitions between presenters.) The higher-echelon presenter would then be the group representative and present the counterarguments (Section 2.4.3.5). In responding to EXCON queries, the presenter whose lane in which the question falls would respond.

This second option (hybrid presentation) would apply in both Step 3 and Step 4. That is, the action approval briefing in Step 3 would be conducted by presenters from both staff layers and delivered to an EXCON representative (or possibly a representative of a still higher echelon), and the engagement presentation would be made by the same two players.

Section 2.2.3.1.2 states that all players for a team must present an action before a player can present a second action; that is, players must rotate presenting so that everyone gets a turn. While this rule still applies in principle in a case where players represent multiple staff layers, it must be applied thoughtfully. Specifically, if the operational staff presents staff mechanics and guidance and the tactical staff presents execution details, turn-taking should not force a player who is part of the tactical staff to present the operational portion of the briefing of vice versa. Rotation should be enforced within the represented staff, not across the whole team.

A two-presenter combined hybrid presentation format will inevitably require slightly more time; adjust accordingly.

5.4.3. Asymmetrical Design: Multiple Layers of Players for BLUE, Single Layer for RED

Many of the interesting dynamics between staff layers are of greatest interest and are easiest to depict in a wargame for the BLUE side. That is, the Marine Corps (or other BLUE) training audience is most interested in how the different layers of the friendly force and interagency are supposed to interact and coordinate to plan and conduct information, and the MCIOC-based (or other) EXCON are best equipped to identify and present those processes with realism and accuracy. While there is certainly value in Marine Corps training audiences understanding and playing from the perspective of an adversary or competitor force, the details of adversary bureaucracy and processes are both harder for EXCON to know in sufficient detail and of less utility to the training audience. Marines with OIE roles will have to understand and interact with multiple staff layers in their future assignments, whereas understanding a specific adversary's approach to coordination is only potentially useful and not as high a priority.

Just because the BLUE side is broken into two or even three layers of staff players does not mean that the RED side needs to be similarly structured, have the same total number of players on its side, or follow exactly the same processes. What is needed is some effort to preserve the reality and appearance of a degree of fairness. Independent of the total number of players on each side and the number of layers either side includes, the game is still fair if the scenario is relatively fairly balanced; if both sides get to present the same number of actions at the same level of detail during Step 4, adjudicated using the same scoresheets; and if both sides are subject to some kind of approval process by which an action might not be approved in Step 3. Any additional activities based on or conducted by additional staff layers on the BLUE side are just additional opportunities for experience and learning, rather than unfair advantages for BLUE.

So, it is a viable option to have two or even three echelons of BLUE represented by participants, while having only a single staff on the RED side, with the contributions of other echelons covered by EXCON RFIs and role-players (and receiving less overall attention).

NOTE: If the teams for the two sides are not symmetrical (that is, if there are more layers being played by BLUE than by RED), some effort will need to be made to assure participants that the game remains fair. The perception of fairness is almost as important as the reality of fairness. The EXCON lead should explain to all players on both sides why one side is playing more staff layers than the other side, how those staff layers are being portrayed for the "thinner" side, and that the scenario and the resources available to each side are still balanced. The explanation might note that, if anything, the side with more played layers is at a disadvantage, as there will unavoidably be more scrutiny and more that can go wrong with those additional layers and that this disadvantage will more than offset the possible advantage from increased total personnel representing the side.

During the playtests and execution of the IWX 21.1 wargame, participants identified RED SMEs to help organize the RED team from both an organizational and process perspective and then advise during game play. This approach presented an additional learning opportunity for all participants but is by no means a requirement for a successful IWX Wargame.

5.4.3.1. Time Cost Associated with Including Multiple Staff Layers as Players

If the game will include representatives from multiple staff layers, additional time will be required to accommodate their interaction:

- If there is an approval/confirmation briefing, time will need to be allocated for it; see Section 2.3.9.

- Inclusion of both an operational- and a tactical-level staff, even if combined as part of the same team, could necessitate additional time for Step 2, perhaps 15–30 more minutes.
- Inclusion of both an operational- and tactical-level staff might affect the way in which Step 3 is conducted but should not change the time required of Step 3.
- Inclusion of both an operational- and tactical-level staff will increase time required for Step 4. Not only will additional time be required to accommodate a wider range of questions and discussion prompted by the EXCON panel of judges, but the individual timed presentation periods will need to be adjusted. Specifically, rather than a single player receiving a single allocation of 5 minutes to present their action (see Section 2.4.3.3), if hybrid presentations are used, more time should be allowed: either a single period of 6 minutes for the two presenters to divide as they please, or two timed periods not to exceed 4 minutes, one for each speaker. The time allocated to rebuttal and counterargument can probably remain unchanged.

5.5. Increasing the Intensity in a Competition Wargame

As noted in Section 5.3, competition scenarios can promote risk-aversion. Although this may be realistic, it can lead to subtle and predictable progress during the wargame and reduced learning opportunities. Advanced or experienced players may fall into predictable patterns of action, whether they are in competition or conflict scenarios. Risk-averse, "easy win" actions may be self-evident, but there are certain cues that EXCON might watch for: If all the actions for one side are scored as "effect is **easy to achieve** with this action" or "magnitude of behavior change sought is **tiny**," this might suggest that risk-aversion has set in. In such cases, EXCON may wish to spice things up a little by increasing the intensity of the IWX Wargame.

Section 2.6.1.2 includes a lengthy discussion of the use of injects within the game as a form of stimulus for the players. EXCON designers for competition scenarios are urged to have a stable of injects available and ready to use.

Another option to stimulate intensity in competition scenarios is to include clear and bold guidance to the players from a higher echelon. Rather than allowing teams to develop plans that contribute only slightly to vague and long-term competition goals, have their forces tasked with some more specific and more aggressive "gray zone" mission (while still remaining below the threshold of conflict) that they must support. This might involve some sort of clandestine activity; a short-term goal to change the behavior of a specific influencer, key figure, or leader; or something else. See Section 5.2 for further discussion.

NOTE: Guidance or objectives that demand more aggressive actions could be in place from the outset in initial scenario guidelines, but they do not have to be. One option for an inject in a later turn would be the arrival of orders with a new objective or orders to pursue an existing objective more fervently.

FIGURE 22

Players Confer in the Engagement Space During IWX 23.3

Photo credit: Stephen Webber, RAND.

Annex. Methods Used in Developing the IWX Wargame

This annex describes the research and other efforts that went into the development and playtesting of this wargame. In a traditional RAND report, the documentation of methods and background would be placed more prominently. However, because the primary purpose of this rulebook is to document the rules themselves, we relegate this (still important!) collection of details to an appendix.

The sponsor for this research, the Marine Corps Information Operation Center (MCIOC), sought RAND's help in developing a wargame for use as part of MCIOC's twice-annual Information Warfighter Exercise (IWX, formerly Combined Unit Exercise, or CUX). Traditionally, the two-week IWX/CUX included a period of "opposed free play" during the second week, in which various groups of participants would expose their plans to a "murder board" of senior peers for critique or face ad hoc adjudication of the effectiveness of their plans versus the effectiveness of a RED team's plans during different phases of a notional scenario operation. Two of the authors had experience with these previous efforts at gamification: Jim McNeive, in his role at MCIOC, and Christopher Paul, during previous RAND work in support of MCIOC (including serving as a senior mentor during several iterations of CUX). While CUX and IWX benefited from the inclusion of a practical planning component with some stimulation of and friction for that plan, these efforts left a significant margin for improvement available as an exercise wargaming experience.

This, then, was the foundation of the requirement MCIOC sought to meet with RAND's support: a better wargame structure for the IWX that had formal elements of wargames, including clearly defined teams, time-constrained turns, defined actions, a structured adjudication process, and a clear set of rules to govern those processes. Coauthor and principal MCIOC point of contact Jim McNeive brought an initial vision of the game: a three-day/six-turn wargame focused on the influence mission of Marine Corps information forces in which two teams would each take a number of actions per turn that would be formally adjudicated using some form of matrix adjudication (described in greater detail below). Working with McNeive and with the support of many MCIOC personnel, the RAND team refined this vision into a codified ruleset, supported the playtesting of the ruleset over several iterations, and produced the initial IWX Wargame in 2021. Over the subsequent two years, the wargame continued to be a prominent part of IWX even as IWX and the wargame were presented in various formats and contexts, for different participants, using a variety of different scenarios. Continued evolution of the wargame led to the publication of a collection of optional supplemental rules for IWX Wargames for operational-level staffs or for competition scenarios later in 2021. Continued experience with the game has led to further evolution and streamlining of the rules; this second edition folds in the rules from the first supplement and captures the state of the art of the rules for the game as used in 2023. This annex documents the development process for the initial ruleset and its subsequent evolution.

A.1. Background: Information as a Joint Function and as a Marine Corps Warfighting Function

Information became a joint function in 2017 and a Marine Corps warfighting function in 2019. Subsequent to the publication of the first edition of this ruleset, the Marine Corps published new doctrine for information: first Marine Corps Doctrinal Publication 8, *Information*, in 2022, and then MCWP 8-10, *Information in Marine Corps Operations*, in 2024.This document has been updated to use terms and concepts consistent with that current doctrine.

While no longer used in Marine Corps–specific doctrine, operations in the information environment (OIE) is a relatively new term of art that emerged first within the 2016 *Department of Defense Strategy for Operations in the Information Environment* and was echoed in the 2018 *Joint Concept for Operating in the Information Environment (JCOIE)*. The Marine Corps has placed strong emphasis on information, designating information as a warfighting function, establishing a Deputy Commandant for information, and creating force structure for information within the MEFs (the MIGs).

Writings predating new Marine Corps doctrine established Marine Corps OIE as having seven functions:[13]

- Assure enterprise command and control (C2) and critical systems.
- Provide information environment battlespace awareness.
- Attack and exploit networks, systems, and information.
- Inform domestic and international audiences.
- Influence foreign TAs.
- Deceive foreign TAs.
- Control OIE capabilities, resources, and activities.

These seven functions still appear as subelements in MCWP 8-10. While all seven of these functions are amenable to treatment within a wargame,[14] the scope on intended training for IWX (and thus for the IWX Wargame) emphasizes "influence foreign target audiences," with a secondary focus on "inform domestic and international audiences" and "deceive foreign target audiences." "Attack and exploit networks, systems, and information" might be included in the wargame, and current rules could cover adjudication of such functions; including the other functions might require some improvisation or adjustments to the rules.

A.1.1. Background: Information Capabilities

One of the key insights of the *Joint Concept for Operating in the Information Environment* is that everything the joint force says or does has the potential to create or affect information, and so the joint force should deliberately leverage "the inherent informational aspects of military activities" as part of information power.[15] In addition to the inherent informational aspects, there are also capabilities that are explicitly and primarily intended to generate, preserve, project, or deny information. Such capabilities have traditionally been referred to as information-related capabilities (IRCs). We note "traditionally," as the latest revision of Joint Publication 3-0, *Joint Operations*, instead describes "joint force capabilities, operations, and activities for leveraging information."[16] Under that heading, JP 3-0 lists the following items: key leader engagement (KLE), public affairs (PA), civil-military operations (CMO), military deception (MILDEC), military information support operations (MISO), operations security (OPSEC), signature management (SIGMAN), electronic warfare (EW), combat camera (COMCAM), historians, space operations, special technical operations (STO), and cyberspace operations (CO).

Marine Corps doctrine now uses the term *information capabilities* to describe categories of information activities: inform, influence, deception, electromagnetic spectrum, cyberspace, space, data fusion and visualization, operations security, and special technical operations. The information activities within these capability categories correspond most closely to what used to be called IRCs. Table 8 lists the information capability categories and the information activities.

This wargame allows that any of the forces that perform any information activities described in Marine Core doctrine, or any of the forces that conduct joint force capabilities, operations, and activities for leveraging information might be made available to either side in a scenario. The focus, however, remains on capabilities to influence, with capabilities to inform or capabilities for deception receiving secondary emphasis.

[13] Eric Schaner, "What Are OIE?" *Marine Corps Gazette*, April 2020, p. 20.

[14] See, for example, Christopher Paul, Yuna Huh Wong, and Elizabeth M. Bartels, *Opportunities for Including the Information Environment in U.S. Marine Corps Wargames*, RAND Corporation, RR-2997-USMC, 2020.

[15] U.S. Joint Chiefs of Staff, 2018, p. 1.

[16] Joint Publication 3-0, 2022, pp. III-17 to III-23.

TABLE 8

The Means of Conducting Information

Information Capability/ Activity Category	Information Activities
Inform	Communication strategy and operations
Influence	Military information support operations
	Key leader engagement
	Civil-military operations
	Civil affairs operations
Deception	Joint military deception
	Tactical deception
	Deception in support of operations security
Electromagnetic Spectrum	Electromagnetic support
	Electromagnetic attack
	Electromagnetic protection
	Electromagnetic spectrum management
Cyberspace	DoD Information Network operations
	Defensive cyberspace operations
	Offensive cyberspace operations
Space	Space domain awareness
	Space control
	Intelligence, surveillance, and reconnaissance
	Position, navigation, and timing
	Satellite communications
	Environmental monitoring
	Missile warning
Data Fusion and Visualization	Information environment battlespace awareness
Operations Security	Signature management
	Physical security
Special Technical Operations	Various activities

SOURCE: Reproduces MCWP 8-10, 2024, Table 1-6.

A.2. Background: Will to Fight

Will to fight is the disposition and decision to fight, act, or persevere when needed. Marine Corps doctrine centers on the idea that will to fight is the single most important factor in war: War is, ultimately, a contest of opposing, independent, and irreconcilable wills.

Understanding an adversary's will to fight can help to identify vulnerabilities to target, as well as provide a critical input to overall operational assessments. If the ultimate objective in war is to get the adversary to stop fighting, and the ultimate objective in competition is to get the enemy to stop competing, then a successful attack on their will to fight may be decisive. Preserving will to fight also requires some self-examination to identify and shore up vulnerabilities or build resilience. Information activities are particularly useful in attacking will to fight.

Previous RAND research developed an analytic model of will to fight that we applied in the development of the IWX Wargame.[17] This model helps the user to identify the most relevant will-to-fight factors of both adversary and friendly units, in that particular instance. It is important to remember that will to fight is dynamic and ever-changing, so an initial baseline assessment can provide useful insights, but it must be updated as the inputs change.

The RAND Will-to-Fight Model includes 29 factors and nine additional contextual factors. The 29 factors range from individual-level training and ideology to unit cohesion and leadership, organizational integrity and support, civil-military relations, and popular support at the national level. Table 9 lists the 29 factors and their associated subfactors.

Assessment of will to fight can be done through tools developed by RAND researchers to support the U.S. military. For the 2020 IWX, the RAND team used these tools to conduct a will-to-fight assessment for both the BLUE and RED forces used in the scenario. These assessments identified critical vulnerabilities that could be exploited by players participating in the IWX.

Results from an effort to attack adversary will to fight should be compared against the baseline will-to-fight assessment conducted prior to the IWX. For example, if the assessment shows (as that one did) a potential weakness in unit leadership and the relationship between unit leaders and marines in one unit, and the RED side attacks that relationship with an information activity, then they should be rewarded with a low score (easier roll) by the adjudicators. If a team fails to use the assessment, or consider will to fight, and attacks an adversary strength, then they should be punished with a high score and target number (more challenging roll).

Will to fight can be attacked via information activities and kinetic activities, and ideally via both. For example, the combination of psychological operations leaflet drops and bombing in the 1991 Gulf War contributed to the surrender of thousands of Iraqi Army soldiers.

Not all successful attacks on will to fight will result in a unit breaking and running. Results are often more nuanced. For example, the unit may hesitate (which can be represented in a wargame as losing a turn of action). The unit may waver in its attacks, affecting its chances of kinetic success. In some cases, loss of will to fight may be temporary and can be recovered over time.

As difficult as tracking will to fight can be in real life, it can be even more challenging in a wargame to provide a realistic snapshot of the will of each side during a wargame. We explored comparative will-to-fight trackers that assigned points to each side in early runs of the IWX Wargame. Although this can provide a helpful visual depiction of each side's status, and more importantly a comparison, we found this method to be overly subjective. Worse, proliferation of such an approach, with a potentially misleading level of confidence, risks misguiding a decisionmaker. Instead, our recommendation is to conduct a baseline assessment and then direct both teams, RED and BLUE, to remain attuned to both adversary and friendly will to fight and to bear in mind the potential value in attacking adversary vulnerabilities and where their own vulnerabilities lie.

For more information on RAND's body of will-to-fight research, see https://www.rand.org/ard/projects/will-to-fight.html. Ben Connable was the lead author of RAND's seminal piece on this topic, which provides a thorough overview: *Will to Fight: Analyzing, Modeling, and Simulating the Will to Fight of Military Units* (2018). For more of an executive summary, see Ben Connable et al., *Will to Fight:*

[17] Connable et al., 2018.

TABLE 9

Factors in the RAND Will-to-Fight Model

Level	Category	Factors	Subfactors
Individual	Individual Motivations	Desperation	
		Revenge	
		Ideology	
		Economics	
		Individual Identity	Personal, Social, Unit, State, Organization, Society
	Individual Capabilities	Quality	Fitness, Resilience, Education, Adaptability, Social Skills, Psychological Traits
		Individual Competence	Skills, Relevance, Sufficiency, Sustainability
Unit	Unit Culture	Unit Cohesion	Social Vertical, Social Horizontal, Task
		Expectation	
		Unit Control	Coercion, Persuasion, Discipline
		Unit Esprit De Corps	
	Unit Capabilities	Unit Competence	Performance, Skills, Training
		Unit Support	Sufficiency, Timeliness
		Unit Leadership	Competence, Character
Organization	Organizational Culture	Organizational Control	Coercion, Persuasion, Discipline
		Organizational Esprit de Corps	
		Organizational Integrity	Corruption and Trust
	Organizational Capabilities	Organizational Training	Capabilities, Relevance, Sufficiency, Sustainment
		Organizational Support	Sufficiency, Timeliness
		Doctrine	Appropriateness, Effectiveness
		Organizational Leadership	Competence, Character
State	State Culture	Civil-Military Relations	Appropriateness, Functionality
		State Integrity	Corruption, Trust
	State Capabilities	State Support	Sufficiency, Timeliness
		State Strategy	Clarity, Effectiveness
		State Leadership	Competence, Character
Society	Societal Culture	Societal Identity	Ideology, Ethnicity, History
		Societal Integrity	Corruption, Trust
	Societal Capabilities	Societal Support	Consistency, Efficiency

SOURCE: Reproduced from Connable et al., 2018.

Returning to the Human Fundamentals of War (2019). More recently, RAND has published a couple of case studies applying the will-to-fight model: Ben Connable, *Iraqi Army Will to Fight: A Will-to-Fight Case Study with Lessons for Western Security Force Assistance* (2022) and Molly Dunigan and Anthony Atler, *Will to Fight of Private Military Actors: Applying Cognitive Maneuver to Russian Private Forces* (2023).

Finally, it is worth noting that will to fight, though challenging to accurately assess, continues to have a huge impact on operational outcomes. In fact, in the Fiscal Year 2023 National Defense Authorization Act, Congress expressed its dissatisfaction with recent will-to-fight assessments (or lack thereof) by directing the Director of National Intelligence to develop a report "examining the extent to which analyses of the military will to fight and the national will to fight informed the all-source analyses of the intelligence community regarding how the armed forces and governments of Ukraine, Afghanistan, and Iraq would perform at key junctures."[18] Clearly, this topic needs additional research and analysis, but the model shown in Table 9 provides a good starting point.

A.3. Background: Importance of Wargaming to the Marine Corps

Both the U.S. Department of Defense (DoD) and the Marine Corps have renewed their interest in wargaming in recent years. The Marine Corps in particular is poised to invest a considerable amount of resources into improving service-level wargaming capability and increasing its number of annual wargames. Wargaming also remains an important tool for operators and for policymakers for learning, exploring, and thinking through potential consequences of planned or conceivable operations. Wargaming is an established tool in the defense and intelligence communities and is especially salient when planners are faced with difficult, complex problems and uncertain futures. For these reasons, resilient inclusion of the information environment and related considerations and effects in wargaming is important for the Marine Corps.

Indicative of a broader, revived interest in wargaming even before some of the more recent high-level focus on wargaming, in the past decade both the Army and Navy war colleges published wargaming handbooks that sought to better describe and instruct on wargaming practice.[19] Other signs of recent, renewed DoD interest in wargaming include high-level memos on wargaming by the Deputy Secretary of Defense and Secretary of the Navy,[20] a new Defense Wargaming Alignment Group (DWAG),[21] the creation of a DoD wargame incentive fund,[22] and DoD-sponsored wargaming conferences.[23] The Military Operations Research Society (MORS) created a wargaming certificate program in 2017 in response to the increased demand for wargamers within the defense community.[24] DoD interest in wargaming has also coincided with renewed interest in wargaming by other countries. The UK Ministry of Defence published its own wargaming handbook in 2017 that, unlike the Army and Navy wargaming handbooks, became doctrine.[25] In 2023 it published another handbook tailored specifically for influence wargaming.[26] China has also invested in computerized wargaming over the past decade, beginning with wargaming strategic problems but also expanding to interservice wargames and tactical simulations.[27]

Within this environment, the Marine Corps is expecting to increase its wargaming capabilities to better prepare for future combat. The Marine Corps not only expects to increase its volume of wargames, but also to increase the technological sophistication of the wargaming it currently conducts.[28] Marine Corps Systems Command is currently overseeing the development of a "world-class" wargaming capability that seeks to be data-enabled and analytically rigorous, incorporating computerized modeling and

[18] Public Law 117-263, James H. Inhofe National Defense Authorization Act for Fiscal Year 2023, 117th Congress, 2021–2022, December 23, 2023.

[19] James Markley, *Strategic Wargaming Series Handbook*, U.S. Army War College, Center for Strategic Leadership and Development, July 2015; and Shawn Burns, *War Gamers' Handbook: A Guide for Professional War Gamers*, U.S. Naval War College, undated.

[20] Bob Work, Deputy Secretary of Defense, "Wargaming and Innovation," memorandum from the Secretary of Defense for Service Principals, February 9, 2015; and Ray Maybus, Secretary of the Navy, "Wargaming," memorandum for Chief of Naval Operations and Commandant of the Marine Corps, May 5, 2015.

[21] Bob Work and Paul Selva, "Revitalizing Wargaming Is Necessary to Be Prepared for Future Wars," *War on the Rocks*, December 8, 2015.

[22] Garrett Heath and Oleg Svet, "Better Wargaming Is Helping the U.S. Military Navigate a Turbulent Era," *Defense One*, August 19, 2018.

[23] Phillip Pournelle, ed., *MORS Wargaming Special Meeting, October 2016, Final Report*, Military Operations Research Society, 2017, p. 5; Phillip Pournelle and Holly Deaton, eds., *MORS Wargaming III Special Meeting, 17–19 October 2017, Final Report, April 2018*, Military Operations Research Society, 2018, p. 2.

[24] Military Operations Research Society, "Certificate in Wargaming," webpage, undated.

[25] United Kingdom Ministry of Defence, Development, Concepts and Doctrine Centre, *Wargaming Handbook*, LCSLS Headquarters and Operations Section, August 2017.

[26] United Kingdom Ministry of Defence, *Influence Wargaming Handbook*, LCSLS Headquarters and Operations Section, July 2023.

[27] Dean Cheng, "The People's Liberation Army on Wargaming," *War on the Rocks*, February 17, 2015.

[28] Todd South, "Marine Wargaming Center Will Help Plan for Future Combat," *Marine Corps Times*, September 19, 2017.

simulation (M&S) and using in-stride game adjudication.[29] Former Commandant of the Marine Corps General Robert Neller, now the namesake for the center, spoke about his desire for a "Star Trek–like holodeck" for wargaming.[30] His efforts ultimately resulted in a multimillion-dollar facility, occupying over 100,000 square feet, capable of wargaming at all classifications for the joint force, on Marine Corps Base Quantico, set to open in 2025.[31] Stakeholders within the Marine Corps wargaming community often express the desire for more sophisticated adjudication, visualization, analysis, and knowledge management for future Marine Corps wargaming.

Marine Corps wargaming constitutes a very broad set of activities. These include training events and simulations, discussion groups and seminars, planning exercises, reviews of plans, and course of action (COA) wargaming as part of the Marine Corps Planning Process (MCPP).[32] Other wargame approaches with a heavier emphasis on adjudication that are currently used by the Marine Corps include matrix games, hex-and-counter games, commercial computer games, commercial board games, and manual games used in combination with M&S and analysis. Wargames are used from the tactical level to the service level and above, with wargames dedicated to informing Marine Corps Title 10 responsibilities to organize, train, and equip the force.[33] Stakeholders involved in wargaming are engaged in activities such as concept development, capabilities development, training and education, science and technology development, operational planning, and others.

Independent of the widespread and institutionalized position that wargaming has in the present-day Marine Corps, when used properly, the method itself can assist in critical thinking, individual and organizational learning, and trial-and-error exploration of new concepts and warfighting approaches without costing lives or materiel. This is particularly true as the Marine Corps continues its transition from over a decade of counterinsurgency operations to other forms of warfare that are markedly different but about which the current generation of marines may have no firsthand knowledge. Wargaming thus has the potential to play an important role particularly where the information environment is concerned, as that is a space where concepts and understanding are nascent and in development but where the chances for experimentation in the real world may be very limited. Although wargames do not prove or "validate" concepts and approaches, they can be used to teach principles, offer perspectives on what does not work, and create additional insights into an issue. They have the potential to raise questions and potential consequences that have not previously occurred to participants.[34]

A.4. Background: Wargame Adjudication

One of the central elements in a wargame is the approach to or mechanism for adjudication. How will the outcomes of actions taken by the players be determined within the wargame scenario? Adjudication is the procedure to impartially resolve the outcome of interactions between sides in a game.[35] The wargaming literature acknowledges three basic kinds of adjudication:

- Free adjudication: The results of interactions are determined by the adjudicators in accordance with their professional judgment and experience.[36] The opposing sides reaching a consensus on the likely outcome of a non-kinetic interaction or engagement is a useful adjudication method in an open or mixed open/closed wargame format.
- Rigid adjudication: The results of interactions are determined according to predetermined

[29] Program Manager Wargaming Capability, Marine Corps Systems Command, "PM Wargaming Capability," September 26, 2018, pp. 1–2.

[30] James Clark, "The U.S. Marine Commandant Wants a 'Star Trek'–Style Holodeck for Wargaming," *National Interest*, September 30, 2017.

[31] U.S. Marine Corps, "Naming of the Marine Corps Wargaming and Analysis Center," press release, August 4, 2023.

[32] MCWP 5-10, 2020, pp. 4-2 to 4-3.

[33] U.S. Marine Corps Warfighting Laboratory Future Directorate, "Title 10 Wargaming," undated.

[34] Burns, undated, pp. 3–4.

[35] Burns, undated, p. 51.

[36] Francis J. McHugh, *U.S. Navy Fundamentals of War Gaming*, Skyhorse Publishing, 2013.

rules, data, and procedures such as combat models or a combat results table.[37]

- Semi-free adjudication: a hybrid approach in which interactions are evaluated by something akin to the rigid method, but the outcomes can be modified or overruled by the lead adjudicator.[38]

The different types of adjudication methods have different strengths and weaknesses, and some are better or more poorly suited to different game purposes, types, and styles. For example, rigid adjudication is all but impossible where the types of actions sides might take are difficult to determine in advance or where the outcomes of such actions are contingent on numerous complex factors and their interactions. Such games are better suited to some form of free adjudication. However, free adjudication can be (or can appear to be) biased by the perspectives of the expert judges. For games involving a discrete number of types of actions or involving combat at a relatively high level of abstraction, a more rigid adjudication method is more appropriate. When determining the outcome between two clashing forces, a list of possible modifiers could be consulted, and then a probabilistic determination (through dice or some other random number generation system) could be made based on a combat resolution matrix developed based on historical combats of that type.

The IWX Wargame uses a hybrid adjudication model that should be considered a form of semi-free adjudication. The process (as described in the rules regarding Step 4) begins with the presentation and description of the action, followed by a rebuttal from the other team, followed by a counterargument. This portion of the adjudication process is typical of what is called a *matrix game* or *matrix adjudication process*, which is a form of free adjudication.[39] However, in this game, the matrix discussion informs a more structured and rigid step in the adjudication process: A panel of judges completes scoresheets, which lead to an assessment of the likelihood of success of the action (the target number), which is then subjected to a probabilistic determination (the outcome roll). Following this semi-rigidity of scoring and rolling dice, the adjudication turns back toward free with the description of that dice-determined outcome described by the narrator in a free format.

A.5. From Initial Vision to Playable Game

Within this context characterized by increased emphasis on the information environment, new concepts related to information as a warfighting function, emerging research related to will to fight, and existing practice related to wargaming, the RAND team sought to transform the sponsor's initial vision into a playable wargame. Note that the authors brought considerable expertise to this effort. Some are hobby wargamers who have also been involved in the design and conduct of defense-related wargames (Connable and Paul). Some have considerable experience with information operations/operations in the information environment and related research (Paul and McNeive), and one served a tour as the information operations officer for Marine Forces Europe and Africa (Welch). Several authors' recent RAND research relates to the topic at hand.[40] These experiences informed refinement and specification from the initial concept developed by McNeive into an initial draft of the rules, which were then further refined through iterative discussions with MCIOC personnel. Once the rules and procedures reached a certain level of ripeness, the process of rehearsal and playtesting began. While the initially developed core concept and structure survived all the rigors of playtesting, playtesting did lead to continued evolution, minor adjustments, and additional optional rules, as well as refinements in the various supporting play aids. The rules reported here include improvements from all playtesting, including the repeated actual execution of the wargame from 2020 through 2023 as part of various forms and instances of IWX.

[37] McHugh, 2013.

[38] McHugh, 2013.

[39] John Curry and Tim Price, *Matrix Games for Modern Wargaming: Developments in Professional and Educational Wargames*, History of Wargaming Project, 2014.

[40] See, for example, Paul, Wong, and Bartels, 2020; and Connable et al., 2018.

A.6. Looking to the Future

As of this writing in late 2023, the IWX Wargame is in its fourth year and is going strong. There is increasing demand for slots in MCIOC-hosted IWXs, and there is a growing number of requests for MCIOC to bring an IWX to various commands and elements in the field. There is also growing interest in the game among organizations outside of the Marine Corps. The logical next step in the development of the game is to provide pre-published and pre-approved scenarios for use with the game. Currently, those wishing to run the game must find or develop their own scenario. By publishing an official IWX Wargame scenario, we hope to decrease barriers to entry for first-time users without in any way constraining the overall freedom of users to employ whatever scenario they wish.

Furthermore, as the numbers of participants, scenarios, and users grow, so will the foundation of feedback about the game. This could lead to additional supplemental rules, additional optional rules, and further discussion of the nuances necessary in the rules to support play related to certain missions or contexts.

A.7. Probability Distribution for Rolling Three Six-Sided Dice

Table 10 shows the probability distribution for rolling and summing three six-sided dice. While initially used during playtesting to help calibrate the scoresheets and the target number formulae, playtest judges found the table to be a useful reference for understanding how likely teams were to succeed at actions with various target numbers. This table is also included in *GA10: Making an Outcome Roll.*

TABLE 10

Probability Distribution for Rolling Three Six-Sided Dice

Target Number	% Chance of Reaching or Beating the Target Number	Target Number	% Chance of Reaching or Beating the Target Number
3	100	11	50
4	99.5	12	37.5
5	98	13	26
6	95	14	16
7	91	15	9
8	84	16	5
9	74	17	2
10	62.5	18	0.5

References

Army Doctrine Publication 1-02, *Terms and Military Symbols*, Headquarters, Department of the Army, August 14, 2018.

Burns, Shawn, *War Gamers' Handbook: A Guide for Professional War Gamers*, U.S. Naval War College, undated.

Cheng, Dean, "The People's Liberation Army on Wargaming," *War on the Rocks*, February 17, 2015.

Clark, James, "The U.S. Marine Commandant Wants a 'Star Trek'-Style Holodeck for Wargaming," *National Interest*, September 30, 2017.

Connable, Ben, *Iraqi Army Will to Fight: A Will-to-Fight Case Study with Lessons for Western Security Force Assistance*, RAND Corporation, RR-A238-1, 2022. As of October 7, 2023: https://www.rand.org/pubs/research_reports/RRA238-1.html

Connable, Ben, Michael J. McNerney, William Marcellino, Aaron Frank, Henry Hargrove, Marek N. Posard, S. Rebecca Zimmerman, Natasha Lander, Jasen J. Castillo, and James Sladden, *Will to Fight: Analyzing, Modeling, and Simulating the Will to Fight of Military Units*, RAND Corporation, RR-2341-A, 2018. As of November 20, 2023: https://www.rand.org/pubs/research_reports/RR2341.html

Connable, Ben, Michael J. McNerney, William Marcellino, Aaron B. Frank, Henry Hargrove, Marek N. Posard, S. Rebecca Zimmerman, Natasha Lander, Jasen J. Castillo, James Sladden, Anika Binnendijk, Elizabeth M. Bartels, Abby Doll, Rachel Tecott, Benjamin J. Fernandes, Niklas Helwig, Giacomo Persi Paoli, Krystyna Marcinek, and Paul Cornish, *Will to Fight: Returning to the Human Fundamentals of War*, RAND Corporation, RB-10040-A, 2019. As of October 7, 2023: https://www.rand.org/pubs/research_briefs/RB10040.html

Curry, John, and Tim Price, *Matrix Games for Modern Wargaming: Developments in Professional and Educational Wargames*, History of Wargaming Project, 2014.

Dunigan, Molly, and Anthony Atler, *Will to Fight of Private Military Actors: Applying Cognitive Maneuver to Russian Private Forces*, RAND Corporation, RR-A355-1, 2023. As of September 29, 2023: https://www.rand.org/pubs/research_reports/RRA355-1.html

Heath, Garrett, and Oleg Svet, "Better Wargaming Is Helping the U.S. Military Navigate a Turbulent Era," *Defense One*, August 19, 2018.

Joint Publication 3-0, *Joint Operations*, U.S. Joint Chiefs of Staff, incorporating change 1, June 18, 2022.

Joint Publication 3-13, *Information Operations*, U.S. Joint Chiefs of Staff, incorporating change 1, November 20, 2014.

Marine Corps Doctrinal Publication 1-4, *Competing*, U.S. Marine Corps, December 14, 2020.

Marine Corps Doctrinal Publication 8, *Information*, U.S. Marine Corps, June 21, 2022.

Marine Corps Warfighting Publication 5-10, *Marine Corps Planning Process*, U.S. Marine Corps, as amended August 10, 2020.

Marine Corps Warfighting Publication 8-10, *Information in Marine Corps Operations*, U.S. Marine Corps, 2024.

Markley, James, *Strategic Wargaming Series Handbook*, U.S. Army War College, Center for Strategic Leadership and Development, July 2015.

Maybus, Ray, Secretary of the Navy, "Wargaming," memorandum for Chief of Naval Operations and Commandant of the Marine Corps, May 5, 2015.

Mazarr, Michael J., Jonathan S. Blake, Abigail Casey, Tim McDonald, Stephanie Pezard, and Michael Spirtas, *Understanding the Emerging Era of International Competition: Theoretical and Historical Perspectives*, RAND Corporation, RR-2726-AF, 2018. As of August 31, 2021: https://www.rand.org/pubs/research_reports/RR2726.html

McHugh, Francis J., *U.S. Navy Fundamentals of War Gaming*, Skyhorse Publishing, 2013.

MCWP—*See* Marine Corps Warfighting Publication.

Military Operations Research Society, "Certificate in Wargaming," webpage, undated. As of May 13, 2021: https://www.mors.org/Events/Certificates/Certificate-in-Wargaming

Morris, Lyle J., Michael J. Mazarr, Jeffrey W. Hornung, Stephanie Pezard, Anika Binnendijk, and Marta Kepe, *Gaining Competitive Advantage in the Gray Zone: Response Options for Coercive Aggression Below the Threshold of Major War*, RAND Corporation, RR-2942-OSD, 2019. As of May 18, 2021: https://www.rand.org/pubs/research_reports/RR2942.html

Paul, Christopher, Ben Connable, Jonathan Welch, Nate Rosenblatt, and Jim McNeive, *The Information Warfighter Exercise Wargame: Rulebook*, RAND Corporation, TL-A495-1, 2021. As of September 2, 2021: https://www.rand.org/pubs/tools/TLA495-1.html

Paul, Christopher, Michael Schwille, Michael Vasseur, Elizabeth M. Bartels, and Ryan Bauer, *The Role of Information in U.S. Concepts for Strategic Competition*, RAND Corporation, RR-A1256-1, 2022. As of March 3, 2024: https://www.rand.org/pubs/research_reports/RRA1256-1.html

Paul, Christopher, Yuna Huh Wong, and Elizabeth M. Bartels, *Opportunities for Including the Information Environment in U.S. Marine Corps Wargames*, RAND Corporation, RR-2997-USMC, 2020. As of December 3, 2020: https://www.rand.org/pubs/research_reports/RR2997.html

Pournelle, Phillip, ed., *MORS Wargaming Special Meeting, October 2016, Final Report*, Military Operations Research Society, 2017.

Pournelle, Phillip, and Holly Deaton, eds., *MORS Wargaming III Special Meeting, 17–19 October 2017, Final Report, April 2018*, Military Operations Research Society, 2018.

Program Manager Wargaming Capability, Marine Corps Systems Command, "PM Wargaming Capability," September 26, 2018.

Public Law 117-263, James H. Inhofe National Defense Authorization Act for Fiscal Year 2023, 117th Congress, 2021–2022, December 23, 2023.

Schaner, Eric, "What Are OIE?" *Marine Corps Gazette*, April 2020.

South, Todd, "Marine Wargaming Center Will Help Plan for Future Combat," *Marine Corps Times*, September 19, 2017.

U.S. Department of Defense, *Department of Defense Strategy for Operations in the Information Environment*, June 2016.

U.S. Joint Chiefs of Staff, *Joint Concept for Operating in the Information Environment (JCOIE)*, Washington, D.C., July 25, 2018.

U.S. Joint Chiefs of Staff, *Joint Concept for Operating in the Information Environment (JCOIE)*, July 25, 2018.

U.S. Marine Corps, "Naming of the Marine Corps Wargaming and Analysis Center," press release, August 4, 2023. As of October 11, 2023:
https://www.marines.mil/News/Press-Releases/Press-Release-Display/Article/3483276/naming-of-the-marine-corps-wargaming-and-analysis-center/

U.S. Marine Corps Warfighting Laboratory Future Directorate, "Title 10 Wargaming," undated.

United Kingdom Ministry of Defence, *Influence Wargaming Handbook*, LCSLS Headquarters and Operations Section, July 2023.

United Kingdom Ministry of Defence, Development, Concepts and Doctrine Centre, *Wargaming Handbook*, LCSLS Headquarters and Operations Section, August 2017.

Work, Bob, "Wargaming and Innovation," memorandum from the Secretary of Defense for Service Principals, February 9, 2015.

Work, Bob, and Paul Selva, "Revitalizing Wargaming Is Necessary to Be Prepared for Future Wars," *War on the Rocks*, December 8, 2015.

Acknowledgments

The development of the Information Warfighter Exercise (IWX) wargame was a collaborative effort. The contribution of the sponsor principal point of contact, Jim McNeive, was so significant that he is rightly included as a co-author. Special thanks to the Marine Corps Information Operations Center's (MCIOC's) Brian Doty, who had a leading role in the planning and execution of the games and playtests for the first edition. Particularly important in conceiving and testing the updated and revised rules were MCIOC's Thomas Schaeffer, Captain Jessica Parham, William Kline, Christopher Cook, and Brandon Patterson. Many other personnel at MCIOC made valuable contributions as playtesters, in designing the scenario and context for the various iterations of the game over the last few years, and as judges and members of exercise control during playtesting and actual execution; there are too many to acknowledge everyone individually, but all are truly appreciated, nonetheless. Thanks to RAND's Devin McCarthy and Stephen Webber for their help identifying elements in the first edition of the rules that were insufficiently clear, and for helping track the changes in section numbering and cross-referencing that came with the second edition's revisions and additions. We further thank Maria Falvo for her administrative support and for helping with formatting for this and other project documentation. We are indebted to Bill Marcellino and Andrew Lotz for their thoughtful and helpful reviews across editions. Fadia Afashe and James Torr helped bring this document into the final published form, no mean feat given that game rules differ in style and substance from what is typical in RAND reports. Santos Vega created the excellent cover design. Thank you, all.

About This Document

This document provides details and documentation related to a series of RAND projects: "MCIOC Information Warfighter Exercise Wargame Support" from fiscal year (FY) 2020, "MCIOC Information Warfighter Exercise Wargame Support: Continuation" from FY 2021, and "Expanding Options for the Information Warfighter Exercise" in FY 2202. The goals of these projects included supporting the Marine Corps Information Operations Center (MCIOC) in the development of a wargame component for the Information Warfighter Exercises (IWXs) and expanding on the rules and guidance for the game once developed.

This document begins from the baseline of the IWX Wargame rules originally published in 2021,[41] folds in the collection of optional rules captured in Supplement 1 to the IWX Wargame, and updates the rules to correspond to the current U.S. Marine Corps lexicon for information and the most recent evolution of the rules themselves.

Numerous handouts and aids for playing the game, as well as a Player's Guide, are available at www.rand.org/t/TLA495-5.

This rulebook should be of interest to those who wish to conduct (or participate in the administration of) an IWX Wargame at MCIOC. The rulebook should also be of interest to anyone wishing to conduct this wargame as part of a different training exercise series or anyone wishing to design a wargame related to operations in the information environment.

The research reported here was completed in October 2023 and underwent security review with the sponsor and the Defense Office of Prepublication and Security Review before public release.

[41] Christopher Paul, Ben Connable, Jonathan Welch, Nate Rosenblatt, and Jim McNeive, *The Information Warfighter Exercise Wargame: Rulebook*, TL-A495-1, RAND Corporation, 2021.

RAND National Security Research Division

This research was sponsored by MCIOC and conducted within the Navy and Marine Forces Program of the RAND National Security Research Division (NSRD), which operates the National Defense Research Institute (NDRI), a federally funded research and development center sponsored by the Office of the Secretary of Defense, the Joint Staff, the Unified Combatant Commands, the Navy, the Marine Corps, the defense agencies, and the defense intelligence enterprise.

For more information on the RAND Navy and Marine Forces Program, see www.rand.org/nsrd/nmf or contact the director (contact information is provided on the webpage).